Comet Hale-Bopp

Find and enjoy the great comet

ROBERT BURNHAM

CAMBRIDGE
UNIVERSITY PRESS

PUBLISHED BY THE PRESS SYNDICATE OF THE UNIVERSITY OF CAMBRIDGE
The Pitt Building, Trumpington Street, Cambridge CB2 1RP, United Kingdom

CAMBRIDGE UNIVERSITY PRESS
The Edinburgh Building, Cambridge CB2 2RU, United Kingdom
40 West 20th Street, New York, NY 10011-4211, USA
10 Stamford Road, Oakleigh, Melbourne 3166, Australia

First published 1997

Printed in the United Kingdom at the University Press, Cambridge

Typeset in Adobe Berkeley 10.25/14.4pt

A catalogue record for this book is available from the British Library

ISBN 0 521 586364 paperback

Contents

INTRODUCTION *The Great Comet of 1997*

FRED BURGER

A bright comet was long overdue when Hyakutake appeared in the spring of 1996. Fred Burger photographed the comet on March 24, 1996 with a 50mm lens and a 2-minute exposure at f/2 on Kodak Royal Gold 1000 film. The comet's tail shows a disconnection knot about halfway between the comet head and the bright star lying in the tail (see chapter 4).

A year and a half ago, two amateur astronomers discovered their first comet — and everyone's about to share in the fun. Comet Hale-Bopp is due to arrive big and bright in our skies very shortly. It will be a sight not to be missed, and I wrote this book to help you enjoy it. Outstanding comets have been rare indeed. Last spring Comet Hyakutake provided a foretaste of what's coming, but its appearance was all too brief. Back in 1985-86, Halley's Comet disappointed most people, despite its fame. For many backyard astronomers, the last really good one was Comet West in 1976. We're overdue!

The book is structured very simply. The first chapter tells how the comet was found, and the next discusses what scientists know about these objects. Chapter 3 is the heart of the book — a detailed observer's diary that explains where to look for the comet and what it will be doing. Making your own photographic record of the comet's visit is the subject of chapter 4, where you'll find detailed advice. And if you have ever hankered to find your own comet, you're like many skywatchers. So chapter 5 offers tips on how to do it. Finally, the appendices provide orbital elements for the comet, a guide to learning more about comets in general, and a glossary.

It's a pleasure to acknowledge those who helped me prepare this book. First, thanks go to Alan Hale and Tom Bopp for finding it and telling how they did it; also for interesting conversations about comet-hunting and amateur observing. Others who discussed topics or reviewed chapters were Don Davis, Alan Dyer, Steve Edberg, Brian Marsden, Alan Stern, and Don Yeomans. Their contributions and suggestions much improved the book, but of course remaining errors are all mine. I also thank Sienna Software for permission to use scenes from *Starry Night*, the best and most convenient sky-visualization software I've ever used.

Finally, I want to thank dear Patricia. For 20 years she's been a truly good sport about being hauled outdoors at all hours of the night to look at one celestial wonder after another. Thank you.

Hales Corners, Wisconsin
November 1996

A Great Comet is Coming

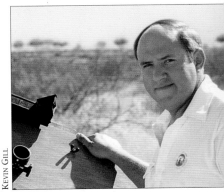

Neither Alan Hale nor Thomas Bopp expected to discover anything that night. Unknown to each other, they were using telescopes 400 miles apart to observe the same star cluster in the constellation of Sagittarius (the Archer). And in less than half an hour on the night of July 22nd to 23rd, 1995, they independently discovered the comet that now bears their names.

Alan Hale spotted the comet first, from his home in Cloudcroft, New Mexico, high in the Sacramento Mountains. For him July 22nd was the first clear evening in over a week. His observing plan that night called for viewing and estimating the brightness of two known comets, Comet Clark and Comet d'Arrest. The 38-year-old Hale is an unusual combination. Founder and director of the Southwest Institute for Space Research, he is a professional astronomer who also has a backyard telescope and strong ties to the amateur community.

By midnight, Hale had logged Comet Clark and was waiting for d'Arrest to rise. His 16-inch Newtonian reflector was set up in the driveway of his house. With over an hour to kill, he decided to explore the star clouds in Sagittarius. Lying due south on July evenings, the Sagittarius region is beloved by amateur astronomers because it is rich in beautiful objects. It has clouds of gas and dust called nebulae, each thousands of times larger than the solar system. It also has sparkling groups of stars called open clusters. While these are similar in kind to the famous Pleiades seen on winter nights, their greater distance makes them look like little clumps of jewels in a telescope's eyepiece. Sagittarius also has globular star clusters, the particular object of Hale's gazing. A typical globular cluster packs several hundred thousand stars into a sphere about 100 light-years across. In a telescope a globular looks like a softly glowing ball of diffuse light.

"There are a *lot* of globulars in Sagittarius," Hale says. "But right next to one globular — M 70 — was a fuzzy object I noticed immediately. I had looked at M 70 two weeks before, but I didn't recall seeing anything beside the cluster."

No celestial object more resembles a comet than a globular cluster. So having spent some 400 unsuccessful hours searching for comets, Hale didn't get too excited. "The chances of pointing a telescope in a random direction and seeing a comet are very slim," he says. "I was thinking I had probably found one of the other globulars in Sagittarius by accident." But upon going into the house and checking with a star atlas, he got a surprise. In that exact piece of sky only M 70 was charted.

"Well," he thought, suspicions now aroused, "if it is a comet it'll soon move." So returning to the telescope he sketched the fuzzy object's position relative to the stars around it and the cluster, and estimated its brightness. If it showed movement, that would be a dead giveaway that the mysterious object was a comet.

Meanwhile, about 400 miles to the west, Phoenix-area amateur astronomer Thomas Bopp had gone into the desert with some friends and their telescopes. They

A first comet for both — Alan Hale (top), a professional astronomer and comet-hunter, had searched for hundreds of hours before he stumbled on the new comet by accident. And Thomas Bopp (above) had no interest whatever in comets before the discovery. He was a deep-sky observer drawn to the soft beauty of galaxies, star clusters, and nebulae. Yet both discovered the comet independently, within about 20 minutes of each other.

were exploring Sagittarius for the same reasons as Hale, but had driven 90 miles to escape the pall of light pollution that covers Phoenix. Bopp, who is in his early fifties, had no prior interest in comets; he was a deep-sky observer — an aficionado of galaxies, clusters, and other "faint fuzzies."

"It was my turn at the scope," Bopp recalls. "I had been watching M 70 drift slowly through the eyepiece of the telescope, which was a 17.5-inch-aperture New-tonian reflector. It's a big scope but it has no motor drive, so you keep it aimed on the stars by hand. Or you can just watch the stars trail across the field of view.

"And that's what I was doing when I noticed a slight glow in the right side of the eyepiece. I asked my friend Jim Stevens, the telescope's owner, if there was another cluster or some other object near M 70."

When Stevens checked the star atlas and said no, Bopp started to get excited. He and Stevens rechecked the charts and arrived at the same answer — no known object. Their words grew animated and caught the attention of the others in the observing party. One of them made a position estimate for the object. Then they all waited an hour to see if it would move.

It did.

A smudge of light betrays Comet Hale-Bopp five weeks after discovery. At the time the comet was more than seven times farther from the Sun than Earth, but already it was brighter than any other comet at that distance. Photo taken September 1, 1995 with the University of Hawaii's 2.2 meter telescope.

DAVID THOLEN & RICHARD WAINSCOAT/UNIVERSITY OF HAWAII–IFA

Certain that they had a comet, the next priority was to report it. But Bopp and his friends were in the middle of nowhere and Bopp's cellular phone didn't work. "In the end," he says, "I jumped in my car and headed back to town."

His difficulties were just beginning. "I got about 20 miles and saw a truck stop. I halted there and tried to send a telegram."

No luck. The Western Union official he spoke to wouldn't send the message because he had never heard of the body that records such discoveries, the International Astronomical Union's Central Bureau for Astronomical Telegrams (see Chapter 5). It is housed at the Smithsonian Astrophysical Observatory in Cambridge, Massachusetts.

Bopp was aghast at Western Union's refusal, and he couldn't believe his ears. Frustrated, he got back in his car and raced home, where he found the exact address for the Central Bureau and called Western Union again. The telegram told of the suspected comet, gave its position, and asked for confirmation.

"And then," says Bopp, "I went to bed. It was about 3 o'clock in the morning."

Back in New Mexico, Hale had his own battle with nerves. After sketching the comet's position and estimating its brightness, about half an hour had passed. "I thought I might have seen it move by this time," he says. "But keep in mind I *wanted* to see it move." While waiting, Hale checked his home astronomy library for possible known objects in the comet's position. And found nothing.

"Well, then," he thought, "maybe it's an older comet having an outburst." So he ran a comet-identification program — and the result was again negative.

"I sent an e-mail to Brian Marsden and Daniel Green at the Central Bureau. I told them I had a possible comet, gave approximate sky coordinates, said that I may have seen some motion, and promised a report once I checked the motion. I logged off, went back outside to the scope, and looked in the eyepiece.

"The comet had moved. No doubt about it.

"I ran back upstairs, logged onto the computer again, and sent Brian another e-mail, saying 'It's moved! It's moving to the west.' And I logged off.

"At that point," grins Hale, "I took my life into my hands. I walked into our bedroom, woke up my wife, and asked if she was interested in taking a look at Comet Hale."

Hale and Bopp had their independent discoveries confirmed the next morning, July 23rd. There was some initial confusion because while both Arizona and New Mexico lie within the Mountain Time Zone, Arizona doesn't adopt daylight saving time. Thus Alan Hale first detected the comet a little after midnight on the 23rd (by his watch, which was on daylight time), while Tom Bopp first saw it around 11:30 p.m. on the 22nd (according to *his* watch, which was on standard time). So Hale was first by less than half an hour.

Comet Hale-Bopp was discovered at an abnormally large distance from the Sun, over seven times Earth's orbital distance. This leads scientists to believe that it is intrinsically big and bright, since most comets are found only when they approach the Sun a lot more closely.

It's coming from above. Comet Hale-Bopp is approaching the Sun from the northern side of the solar system in an orbit that stands perpendicular to the orbits of the planets. After closest approach to the Sun (April 1, 1997), the comet dives southward. Hale-Bopp will next revisit the inner solar system in about 2,400 years. The diagram displays the comet and inner planets on April 1st, when Hale-Bopp lies 125 million miles (202 million km) from Earth.

Exactly how big and bright are unknown. The difficulty comes from uncertainty over the comet's physical size and how active it is. Aside from Comet Halley, photographed by the Giotto spacecraft a decade ago, no one has ever seen a comet close up. How large is Hale-Bopp? Because the comet was unusually bright at discovery, then either it is physically larger than most comets, or else extremely active, responding vigorously to the merest touch of solar warmth. Weighing the possibilities, planetary scientists think Hale-Bopp both has a larger nucleus than Halley and is more active. Their best guess at present gives it a diameter of about 25 miles (40 km), and the comet seems to be quite productive of gas and dust.

Scientists also suspect that Comet Hale-Bopp may prove to be one of the rare "great comets." These are big and beautiful comets that appear about once a century. There is no official definition, but great comets are bright enough to be seen in broad daylight. The Great Comet of 1910 was one such earlier this century. It appeared in January 1910 and was visible to the naked eye when only a few degrees away from the Sun. (In fact, this comet much outshone the apparition of Halley later that same year.)

But one comet in particular haunts Hale-Bopp: Comet Kohoutek of 1973-74. Like Hale-Bopp, Kohoutek also was discovered at a great distance from the Sun. This led astronomers at the time to predict that it would be unusually bright. Carried away by what looked to be a sure thing, astronomers gave it a big build-up

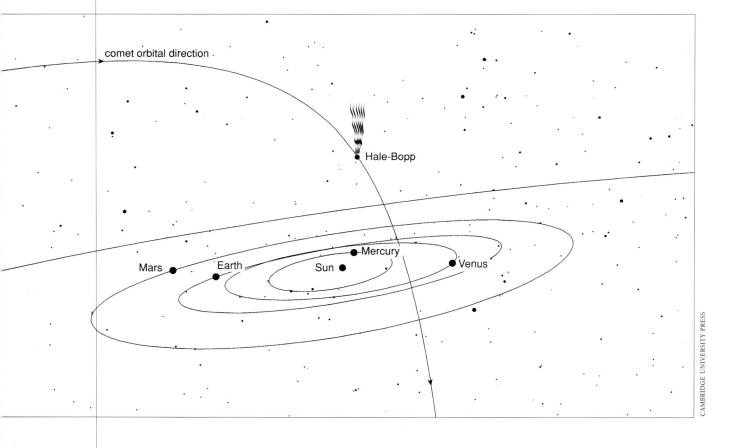

comet orbital direction

Hale-Bopp

Mars   Earth   Sun   Mercury   Venus

CAMBRIDGE UNIVERSITY PRESS

of publicity, attracting world-wide attention. But when the comet finally arrived, it was disappointingly dim. For scientists, Kohoutek's apparition was actually quite successful and much was learned about how comets behave. But the public felt a huge letdown.

After nearly a quarter-century's study of Kohoutek, scientists think they understand why it performed as it did — and why Hale-Bopp won't be a repeat. The answer, surprisingly simple, is that Kohoutek was making its first-ever appearance in the inner solar system. As a so-called "Oort Cloud comet" (see Chapter 2), Kohoutek's anomalous brightness came from a thin layer of condensed frost coating its nucleus. When this layer felt the Sun's warmth for the first time in eons, it burst into bright flourescence and fooled everyone. But once the frost had dissipated, only a relatively inactive nucleus was left.

Comet Hale-Bopp is no such first-time visitor. With an orbital period of a few thousand years, it has made many previous trips around the Sun. It really *is* a large comet and an active one. And there's every reason to expect that it will perform outstandingly as the Great Comet of 1997.

Coming attractions. Comet Hale-Bopp had a forerunner in March and April 1996, when Comet Hyakutake appeared in the evening sky. Scientists are now predicting that Hale-Bopp will outshine Hyukatake. This photo captures Comet Hyakutake on March 23, 1996. Tim Printy used a 16-minute exposure on Fuji SuperG 800 Plus print film with a 135mm lens at f/2.8; the camera rode piggyback on a telescope mounting hand-guided to track the comet.

CHAPTER 2

Icy Visitors From the Depths of Night

Like other comets, Hale-Bopp isn't much physically, at least compared to objects like Jupiter, Mars, or even the Moon. A comet's tail may extend for millions of miles, but its nucleus — the tiny source of all that wispy beauty — measures only a handful of miles across. More fragile than styrofoam, comet nuclei are "dirty snowballs," a term coined by planetary scientist Fred Whipple in 1950. In his icy-conglomerate model, still the best available, the heart of a comet is a lump of water ice laden with silicate and hydrocarbon impurities and covered with a dark, porous crust.

Every bright comet is thus a gaudy act put on for a few weeks or months while the comet is near the Sun. The natural domain for most comets, however, is the darkness that reaches halfway to the nearest stars — a chilly, perpetual twilight zone lit only by the bright arc-light of the Sun and the combined glow of the Galaxy's stars. Out there, comets exist in a kind of deep-freeze combined with a tanning bed. A comet coasts through space, chilled to a temperature not much above absolute zero. Ultraviolet light from the distant Sun and surrounding stars slowly curdles the ices at its surface, transforming them and hydrocarbon compounds into a tarry crust. Over eons, the heat of stars drifting through the Sun's vicinity gently roasts comets, while energetic cosmic rays prickle their surfaces. In

Streamers coming from Hale-Bopp's coma reveal the existence of jets on its unseen nucleus, in this image taken on September 29, 1996. The image, made with a home-built CCD electronic camera attached to a 6-inch Newtonian reflector telescope, has been false-color processed to show the comet's jets.

RICHARD BERRY

time, a rind or crust many meters thick develops on the nucleus, topped by a light layer of volatile frost. For most comets, this is life-everlasting. But for some, random motions or the gravity of a nearby star may nudge the comet out of its placid stasis. It starts the long fall toward the Sun, which is so distant that sunlight ages weeks on its journey to the comet.

Once among the planets, gravitational encounters (especially with Jupiter) may send the comet back into deep space or keep it in orbit close to the Sun. Astronomers draw an arbitrary distinction between comets with long periods (greater than 200 years) and those with short (under 200 years). Comet Halley, with a period of 76 years, is a short-period comet. But Hale-Bopp certainly qualifies for the long-period category: it last passed through the inner solar system some 4,200 years ago and it will be back again in just under 2,400 years. The change comes mostly from perturbations in its orbit by planets.

Few comets have been visited by spacecraft and of these only Halley's has had its nucleus photographed. When Halley returned in 1985-86, the European Space Agency's probe Giotto was part of a fleet to visit it. The probe couldn't land, but it carried a camera. Encountering the comet at a relative speed of 42 miles per second (68 kilometers per second), Giotto passed 370 miles (596 km) from the nucleus.

Giotto's photographs revealed remarkable details. Halley's nucleus is an irregular oblong measuring 9.5 miles by 4.5 miles by 4.5 miles (15.3 km by 7.2 km by 7.2 km). It is tumbling in a complicated way: around one axis it spins every 3.7 days, while it nods slowly once every 7.1 days around another axis. The surface is lumpy and cratered. Sunlight heats the crust to room temperature or hotter, while just below the crust the comet's interior is hundreds of degrees below zero. Where heat penetrates the crust's thin spots, the underlying ices erupt in geysers of dusty gas; Giotto saw three jets in action. As the nucleus rotates, jets turn on when the Sun rises over them and shut down as the Sun sets. They can also act as feeble rocket thrusters and change a comet's orbit.

Extremely dark, Halley's crust reflects just 3% of the light falling on it (roughly like fresh tar), and its structure is granular, brittle, and rubbly. The crust contains silicate dust and hydrocarbon molecules from the comet's impurities — debris left behind by the escaping gases. If you could walk on the comet, its surface would look like a nasty lump of crusty black goo. It would smell very bad and dirty your fingers to touch it; some of its materials are deadly poisonous. Also, radar measurements of Comet Hyakutake showed that its nucleus was shrouded by a blizzard of particles ranging in size from an inch or two across to small boulders. These were thrown off by gas fleeing the nucleus through the porous crust.

In the jets the gas shoots out at about a kilometer per second. It carries off dust and forms a bright shell called the coma, which hides the nucleus from view. Depending on a comet's activity and its distance from the Sun, the coma can extend many thousands of miles in diameter. Like the atmosphere around a planet, the coma is densest at the surface. With water ice a comet's main ingredient, the gas in the coma derives mainly from byproducts created when sunlight breaks up water vapor and the impurities in the ice. Scientists have identified in comets a rich stew of substances, such as carbon monoxide, cyanogen, molecular carbon, methane, ethane, ammonia, formaldehyde, hydrogen cyanide, and methyl alcohol.

After creating the coma, sunlight continues to work the comet over. Photons push dust particles — many about as large as those in cigarette smoke — out of the coma and trail them behind the comet in its orbit. Reflecting sunlight, the dust tail has a white or yellowish color in photos. It usually displays a fairly smooth texture and often curves. Some comets show more than one dust tail and some are streaky. On occasion, a dust tail appears to extend sunward, but this "anti-tail" is an optical illusion caused by the viewing angle.

The gas from the nucleus gets an even rougher ride. First, solar ultraviolet light ionizes it, knocking electrons out of its atoms and molecules and leaving them with an electrical charge. The solar wind, a flow of mostly protons and electrons from the Sun, blows past a comet with a speed of 300 miles per second (500 km/sec) at Earth's orbit. The magnetic fields in the solar wind largely ignore the uncharged dust particles, but quickly snatch up the ionized gas, squeezing and reshaping it into a gas tail. Compared to the dust tail, a gas tail is typically longer, narrower, and bluish in color with ionized carbon monoxide. The gas tail always points away from the Sun, even when the comet is receding from the Sun. The solar wind is turbulent and the comet is charging across its outward flow. This means ionized gas tails often

What a comet nucleus looks like remained largely guess-work until 1986, when the Giotto spacecraft flew by Comet Halley at close range. Halley's nucleus was as black as tar, measured about 10 miles (16 km) end to end, and displayed three bright jets of dusty gas. This irregular body was no larger than a small city, but it was the source for a bright tail that stretched millions of miles.

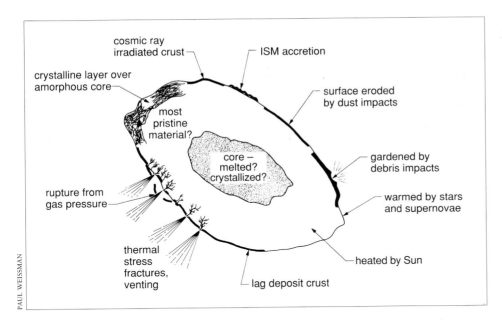

cosmic ray
irradiated crust

ISM accretion

crystalline layer over
amorphous core

surface eroded
by dust impacts

most
pristine
material?

core —
melted?
crystallized?

gardened by
debris impacts

rupture from
gas pressure

warmed by stars
and supernovae

thermal
stress
fractures,
venting

heated by Sun

lag deposit crust

PAUL WEISSMAN

If you chopped open a comet nucleus, its "dirty snowball" nature would be evident. Although the interiors of comets have changed little since the solar system formed 4.6 billion years ago, scientists believe they develop thick crusts of altered organic ices mixed with porous grains of silicate rock dust. As sunlight warms a comet nucleus, its ices evaporate, carrying off dust and forming the coma and the long, streaming tail.

show abrupt changes in direction, tangled knots, and torn-off pieces, much like a plume of chimney smoke in a gale of wind.

Each trip around the Sun brings a comet's demise a little closer. Subjected to solar warmth, a comet's life begins to run out quickly. Scientists think that something like 3 feet (a meter) of comet evaporates on each passage, with up to 6 meters eroding from an active region. For most comets, this means they are good for only about 100 to 1000 apparitions. When the ice is gone, only a loosely cohesive body of dust will be left or, in some cases, a rocky body. Scientists believe an asteroid named 3200 Phaethon is one such burned-out comet. Phaethon shows no comet activity, but its orbit matches that of the Geminid meteor shower. Most scientists believe the shower's meteor particles are comet-dust that was shed ages ago when Phaethon was active.

Seeing a bright comet, people marvel over its beauty. What they don't often realize is that while it appears fresh, a comet is spookily primeval. Comets have been part of the solar system since its infancy and are among its least altered objects. They offer unique insights into the Sun's earliest days.

The solar system began, scientists think, about 4.6 billion years ago when a gigantic cloud of interstellar dust and gas began to collapse. The cloud came from remnants of the early universe mixed with castoff fragments of dead and dying stars. The densest part of the cloud grew denser by pulling in surrounding matter. It steadily grew hotter and began thermonuclear reactions in its core. Thus was born the proto-Sun. Around the nascent star, the leftover cloud soon settled into a broad, thin disk of dust and gas, wheeling slowly. Scientists call this the solar nebula. Hot where it touched the proto-Sun but frigid farther out, the nebula evolved toward today's planetary system. Small particles collided and stuck. These

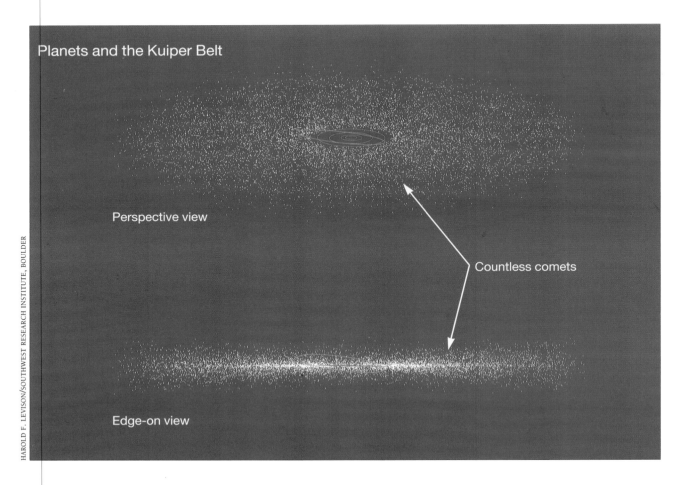

Planets and the Kuiper Belt

Perspective view

Countless comets

Edge-on view

Surrounding the planetary system like a flat collar, the Kuiper Belt forms the "reservoir" and source for short-period comets like Halley's. It extends from about the orbit of Neptune to some 1,000 times farther from the Sun than Earth; this schematic diagram shows only its innermost part. Planetary orbits from Jupiter to Neptune are in red; the eccentric orbit of Pluto (once a member of the Kuiper Belt) is shown in green.

attracted more particles, growing into planetesimals. The planetesimals close to the newborn Sun were too hot to keep their ices and other easily evaporated elements. So they evolved into the rocky terrestrial planets, such as Earth and Mars.

Farther out, where the disk was cooler, gas-giant planets like Jupiter and Saturn formed around rocky cores perhaps ten times the size of Earth. They snatched from the nebula volatile elements unable to survive closer to the Sun, mainly hydrogen and helium. Finally, outside the realm of the main planets, icy "cometesimals" took shape — small, very cold, and loaded with water ice and other volatile substances. Unlike the planets, these bodies did not coalesce into vastly larger objects. Some collided with the planets, contributing water and other elements to them. But many remained on the fringe.

About 50 years ago, astronomers began to realize that comets originated in two reservoirs that lie outside the zone of the planets. The main source for short-period comets is the inner reservoir, called the Kuiper Belt. It is named for Gerard Kuiper, the Dutch-American astronomer who proposed it. The Kuiper Belt extends from about the orbit of Neptune, 35 astronomical units (AU) from the Sun, out to roughly 1,000 AU. (Earth orbits the Sun at an average distance of 1 AU.) Like the original solar nebula it descended from, the Kuiper Belt is flat and disklike. It merges into the Oort Cloud, which reaches to perhaps 100,000 AU or roughly 2 light-years. The Oort Cloud is named for Jan Oort, a Dutch astronomer who discovered its probable existence from studying the orbits of long-period comets. The inner Oort Cloud shares the disklike shape of the Kuiper Belt, but in its outer portions, the Cloud flares out to form a large sphere reaching about halfway to the nearest stars.

New research is showing how this structure has evolved since the solar system's birth. Largest of the planets, Jupiter is the main pot-stirrer. It stopped the formation of a planet between it and Mars, leaving a belt of small rocky asteroids. And Jupiter took comets that came its way and tossed most of them outward, where they formed the Oort Cloud or escaped to interstellar space. Almost all comets passing through the inner solar system today come from the Kuiper Belt. Scientists have recently discovered that under the right conditions, Neptune can grab comets from Kuiper Belt and send them in to where Uranus can perturb their orbits further. From Uranus, they can be handed off to Saturn and thence to Jupiter. At Jupiter they will either be caught in the inner solar system or flung out again to the Kuiper Belt, the Oort Cloud, or out of the system entirely.

In the past five years, astronomers have begun to discover individual cometary bodies in the inner Kuiper Belt. Some 40 of these "transneptunian objects" have been found so far, each about ten times the size of comets that come near Earth. Using the Hubble Space Telescope, a team of researchers believes they have detected

Out of the cold and the dark. In the last five years, planetary scientists have begun to discover icy bodies in the inner Kuiper Belt. One such "transneptunian object" is 1995 QY9, shown boxed in two photos taken a couple hours apart. These objects are thought to have diameters roughly ten times larger than the nuclei of comets like Halley's and Hale-Bopp. Some have orbits that may one day bring them into the inner solar system.

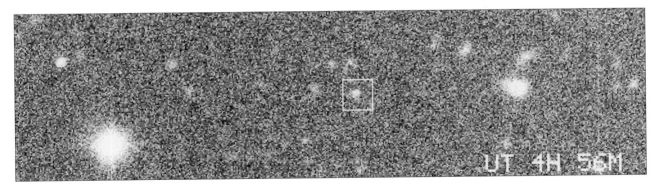

UT 4H 56M

UT 7H 45M

DAVID JEWITT & JANE LUU/UNIVERSITY OF HAWAII

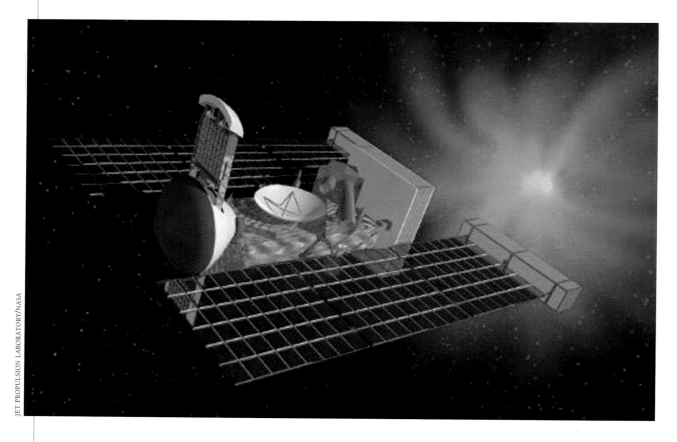

What's next for comet science? NASA is planning a mission named Stardust to go to Comet Wild 2. After rendezvous with the comet in 2004, the spacecraft will fly within 100 kilometers (60 miles) of the nucleus and collect microscopic samples of the comet's gas and dust tails, bringing them back to Earth for analysis.

roughly 40 more objects, somewhat smaller in size. All these objects represent relatively pristine comets which planetary scientists want to study from observatories on the ground and in orbit, and by using a new generation of space missions.

The next phase of comet exploration involves returning samples to Earth. Two missions are well along in design, one from NASA, the other from the European Space Agency. NASA's mission is named Stardust, and it is scheduled for launch in 1999. After a five-year cruise, it will pass through the tail of Comet Wild 2, imaging its nucleus and collecting samples of the gas and dust tails. The samples will be brought back to Earth for lab analysis in 2006.

The European mission is more ambitious. The main craft is called Rosetta and it carries a lander named RoLand. Rosetta is due for launch in 2003 and arrives at Comet Wirtanen in 2011. At arrival the comet will still be far from the Sun. Rosetta will match the comet's orbit and fly in formation with it, releasing the lander onto the comet's nucleus. Rosetta will accompany the comet as it falls toward the Sun, swings around it, and starts to head out again. The craft will spend at least a year studying the comet at close range.

In the next decade or so, the results of these probes (and others under consideration) should produce a marvelously new and detailed view of these icy, ancient visitors from the edge of the Sun's family.

Chapter 3

Comet-Watcher's Logbook

At the start of 1997, Comet Hale-Bopp passes a watershed. The comet's long approach has given amateur and professional astronomers much opportunity to follow its growing activity. But this period of semi-ownership is about to end, as Comet Hale-Bopp moves into the ken of the general public. This chapter provides, first, an outline calendar of what's coming in the next year for observers in the northern and southern hemispheres. Then it tells how to observe a comet with everything from the unaided eye to sophisticated telescopes. And finally it details what you can see on a monthly, weekly, and occasionally daily basis.

COMET CALENDAR — NORTHERN HEMISPHERE

In early **January 1997**, Comet Hale-Bopp will be visible to the naked eye in the western evening sky as a fuzzy patch of light in the faint constellation of Serpens Cauda. It lies below and to the right of the 1st-magnitude star Altair in Aquila. Altair, together with Deneb in Cygnus and Vega in Lyra form the "Summer Triangle," a prominent trio of 1st-magnitude stars. High in the sky during summer nights, these stars are now setting in the west. You can actually see Comet Hale-Bopp both after sunset *and* before sunrise, all on the same night. (This oddity occurs because the Sun lies so far south in the sky — thanks to the winter solstice on December 21st — that it both sets before the comet does and rises after the comet rises.) The result is that you can view the comet low in the west in evening twilight and then get up the next morning and spot it in the east just before daybreak.

But as January advances, predawn views of the comet improve while evening ones worsen, a pattern that holds throughout January and February 1997. During this time, Hale-Bopp rapidly brightens and slowly moves day by day into the Summer Triangle. Seen in the frigid predawn darkness, the comet passes above Altair in late January and by the end of February, stands below Deneb, with its tail reaching up toward the star. By this time the comet will be a memorable sight for early risers.

As **March** commences, the comet remains a morning object, but not for long. Its path carries it to the northeast on the sky away from Cygnus toward Andromeda. After about mid-March, you can stop setting the alarm clock for the wee hours: best views of the comet hereafter come in the evening. Around the third week of March, Hale-Bopp sails past the Great Galaxy in Andromeda (M 31). This is when the comet will be brightest and most visible. Even from urban locations, Hale-Bopp will be impressive, and from darker sites it will be unforgettable. The comet and its tail will be like a pale upright torch in the northwest each evening.

On **April** 1st, the comet passes perihelion, its point of closest approach to the Sun. It will lie at a distance of 85 million miles from the Sun and 125 million miles (137 million km and 202 million km) from Earth. During the first week of April, the comet moves into Perseus, making a beeline toward the Pleiades star cluster in

Previous page. **Sunlight helps cook a rich stew of compounds in comets, including acetylene, methanol, cyanogen, and carbon monoxide, and these provide scientists with clues to the ingredients of the early solar system. In Hyakutake (seen here) scientists detected for the first time methane and ethane, long predicted to occur in comets. (This image is a 4-minute exposure on Fuji SuperG 800 Plus print film using a Ceravolo 190mm f/2.3 Maksutov-Newtonian astrograph.)**

Taurus. The comet passes above the Pleiades during the last week of April, while they and the Hyades cluster sink into the western sky. As each evening finds Comet Hale-Bopp a little lower, the comet cuts across the horns of Taurus, moving towards Orion. The comet is now on its outward path from both the Sun and Earth, and the brightest part of its apparition will probably be over. Finally, soon after the middle of May, evening twilight will swallow the comet and Orion both.

A long hiatus follows for northern hemisphere observers as the comet moves across Orion and Monoceros into Canis Major and Puppis. These constellations and the comet remain lost in the Sun's glare until early autumn. Late in **September**, Comet Hale-Bopp — much dimished in brightness — will briefly reappear low in the morning twilight. Unfortunately, while it was hidden in sunlight, the comet traveled far to the south and when viewers recover it, it will be in rapid retreat toward their southern horizon. From mid-northern latitudes observers will lose sight of it forever sometime late in October, a bare month at best after recovery. The comet's southward course poses a smaller problem for viewers at lower northern latitudes. At latitude 30° north, for example, the comet returns to view in the predawn sky in late August and lingers until the middle of **November**. But eventually even these observers must say farewell to the comet, too.

COMET CALENDAR — SOUTHERN HEMISPHERE

For southern hemisphere observers, the period when the comet is brightest falls locally during late summer and early autumn. This means that the Sun rises before the comet does and sets after it — unfortunately, the exact inverse of the northern hemisphere's favorable apparition. Southern latitudes get their innings, however, after the comet's path turns southward. By the last week of April 1997, southern viewers will start to see the comet in the evening twilight, low in the northwest below the stars of Orion.

During **May**, as southern autumn advances to winter, the comet will climb higher each evening, aiming for the reddish star Betelgeuse in Orion. Hale-Bopp passes just north of the star in the first days of June. The comet's visibility is helped by the southern Sun's descent into winter, with sunsets coming earlier each day. But the inevitable can't be staved off forever. Sometime in late June, Comet Hale-Bopp will vanish into the sunset twilight.

For observers equipped with telescopes, however, the show has by no means ended. The comet returns to view in the predawn sky early in July. While southern winter makes its slow turn toward spring, the comet courses in the general direction of Sirius, the brightest star in the sky. By the end of August, Hale-Bopp lies in Puppis. It spends all September and half October in the old constellation of Argo, Jason's ship. By this time it will likely be around 5th magnitude — dim, but a worthwhile sight in a telescope. Hale-Bopp crosses Puppis, a little corner of Vela, and then settles into Carina to finish out 1997. On **New Year's Eve**, it lies 12° — one handspan — due south of Canopus, the sky's second-brightest star.

HOW TO FOLLOW THE COMET

Bright comets can be seen with any kind of optical aid — or none. What's best to use depends on how bright the comet is and what aspect you want to see. In a moment we'll cover how to view a comet, but at the beginning it's important to correct a couple of common misperceptions.

Despite what television programs sometimes show, comets don't flash across the sky at some special time of night. Comets do move swiftly, of course, but given the distances their apparent motion is quite slow. From one night to the next, a comet's visual location among the stars will not change much. If, for example, you see the comet near Altair at 6 a.m. on a Tuesday, it won't have moved far by the following morning.

Second, comets are not especially colorful or gaudy, at least by comparison with the artificial lights around us. Regrettably, this means that from a typical city or populous suburb, even a relatively bright comet can be hard to see clearly. This won't likely be a problem when Hale-Bopp is at its brightest in March and April 1997, but before or after then the comet won't be super-prominent. (The sky views in this book, however, should help you locate the comet without difficulty during the times when you'll need a little assistance.)

Some people worry that if they have never seen a comet, they might not recognize Hale-Bopp. Don't fret — comets are pretty distinctive! When dim, a comet typically resembles a small fuzzy patch of light. It may show a brighter core and be elongated in some direction, the hallmark of a tail. From late January 1997 onward, Hale-Bopp will be quite remarkable and unlike anything else in the sky. We'll go over details below, but basically you can expect to see a white or bluish streak of light coming from a small head that may look like a bright star surrounded by a halo. (The halo is called the coma.) If Hale-Bopp behaves like Comet Hyakutake did in March and April 1996, it will have a head several degrees across and a straight tail many more degrees long. (Your fingertip held at arm's length covers about 2°.)

City-dwellers and suburbanites sometimes find even a bright comet a bit of a letdown, especially if their expectations derive from television or movies. Comets and other natural wonders of the night sky have a hard time competing with the blaze of artificial lighting. Please don't misunderstand; Hale-Bopp will be an outstanding sight, even for city-dwellers viewing it from their rooftops. The same goes for observers in the suburbs. But the people who will get the most out of this comet — those for whom it will be a spectacle they'll cherish the memory of for years — will be those who see it from away from city lights. Traveling only a dozen miles past the last streetlights gives the comet a big boost in visibility, and seeing it on a moonless night from the countryside will leave you spellbound.

Sound appealing? You could just hop in the car and go, but you'll have more fun if you take a few moments to prepare. Bring this book and a detailed road map. Also take a flashlight, but cover its lens with red tissue paper. (This lets you consult the book and map without ruining your night vision.) Bring warm clothes — and don't

Sometimes the unaided eye gives the best view of a comet, and many observers find that nothing beats eyesight for tracing a comet's tail. Comet Hyakutake's ion tail cuts across the handle of the Big Dipper and keeps on going.

forget bug repellant if your location and the season call for it. A thermos of coffee is also a good idea. Outside town, find a side road with little traffic and a wide view of the sky. Pull completely off the pavement, dousing the headlights. Pick a place where passing cars won't suddenly come upon you with little warning. When you look up, you'll be amazed at how many stars you can see. Look around to get your bearings. Can you see the Big Dipper? Maybe Orion? Or the Milky Way? Give your eyes about 15 minutes to adjust to the dark. Now — do you see the comet? When you locate it, can you tell how big it is, or how long the tail stretches? For a "sky ruler," use the palm of your hand held at arm's length. It spans about 10° (and recall your forefinger covers about 2°).

Naked-eye viewing. Many novice sky watchers are surprised to discover that telescopes aren't always best for viewing some celestial sights. Part of the wonder of a comet is that you can see it with everything from the unaided eye to the biggest telescope imaginable. For example, your eyesight is unsurpassed for tracing the comet's tail. Don't be surprised if you find, as your eyes adapt to the darkness, that the tail just keeps getting longer! It also helps not to look directly at something faint, but a little to one side. Astronomers call this "averted vision," and it works because your peripheral vision is more light-sensitive than that in the center of the retina. Look carefully at the comet. Can you see color? Comet Hyakutake had a distinct bluish or turquoise color in its head. This came from carbon monoxide and molecular carbon emission stimulated to glow by ultraviolet light from the Sun. No one knows just what hues Hale-Bopp will display, so look for color and note whatever you see.

Binoculars. Amateur and professional astronomers like 7x50 binoculars because they combine excellent light grasp and ease of handling. But 7x50s aren't essential; if you have any binoculars, bring them. Binoculars can reveal details in the comet's head and tail that are hard to see with the unaided eye. And sweeping across the comet from a constellation crowded with pinpoint stars is a breathtaking experience. Spotting scopes are also useful, although their higher magnifcations involve a tradeoff. They may show more detail, but the field of view is diminished. Also, you have to use them on a tripod, and Earth's rotation will soon carry the comet (and stars, too) out of the eyepiece's field.

Astronomical telescopes. These vary from basic reflectors on simple mountings that you aim by gentle nudging all the way to sophisticated catadioptric telescopes with built-in computers to let you search-and-find with the click of a control paddle. The variety of instruments on the market is enormous and to a novice, quite bewildering. This isn't the place for a how-to guide on buying a telescope, a process with some pitfalls that aren't obvious to beginners. I'll just say that if you're buying your first telescope to see Comet Hale-Bopp, be very careful. Keep in mind that those "See 1000x!!" telescopes sold in camera and department stores are mostly junk. My recommendation? Before you spend so much as a penny on a telescope, check out the annual buyer's guides from ASTRONOMY and *Sky & Telescope*. These

No telescope? Don't worry — even small binoculars will provide an enhanced view of the head and tail. While amateur astronomers usually prefer 7x50s, any kind are better than none. (Comet Hyakutake on March 22, 1996.)

Over page. Comets typically show two tails, a bluish one of ionized gas and a white or yellowish one of microscopic dust. Both tails originate in material shed by the comet's nucleus, but different forces act upon them. The pressure of sunlight drives dust particles into a broad, fan-like tail, while electrons and protons in the turbulent solar wind ionize the comet's gas, creating an often-contorted, narrow ion tail. (Comet West, photographed in March 1976.)

are available on newsstands and by mail order from the magazines (see the resource guide in the back). Either guide will help you make a smart purchase. Plus they have large listings of what's on the market and much else besides. At less than $5 each, you won't go wrong — especially considering a decent telescope costs several hundred dollars at a minimum.

What to look for in the comet. As Chapter 2 explains, comets are "dirty snowballs" that turn on and spout gas and dust when the Sun warms them. They produce activity at all scales from the huge tail that can stretch for millions of miles down to small luminous jets that come out of the nucleus. Let's start by looking at the tail, an easy sight with the naked eye or in binoculars.

As we've seen, comets typically produce two tails, one of dust, the other of ionized gas. A comet's dust tail shines by reflecting sunlight and appears smoth and white or yellowish. Many are relatively short and some show streaks. Hale-Bopp has been actively producing dust, so its dust tail should be quite a sight. The other tail to look for is the gas tail. Compared to the dust tail, the ionized gas tail is typically longer, straighter, and bluish in color. It may be the brighter. You won't see it move visibly,

DAVID CHURCHILL

Even low magnifications on a telescope helps bring a comet into view. You may not see by eye all the structure in the tail recorded in this photo of Hyakutake, but try anyway. Experienced observers learn to look to one side of the area of interest, detecting faint details with their peripheral vision.

BRAD WALLIS & ROBERT PROVIN

As you increase magnification, details in the coma begin to emerge. You'll see luminous shells and fans which reflect eruptions on the tiny nucleus, buried deep in the brightest part of the coma. (Comet Hyakutake on March 24, 1996.)

Even the great power of the
Hubble Space Telescope can
penetrate the coma only so far.
Here we see the result of an
outburst from the nucleus of
Hale-Bopp — note the curving
bar, produced when the rotating nucleus turns away from
the debris it has ejected.

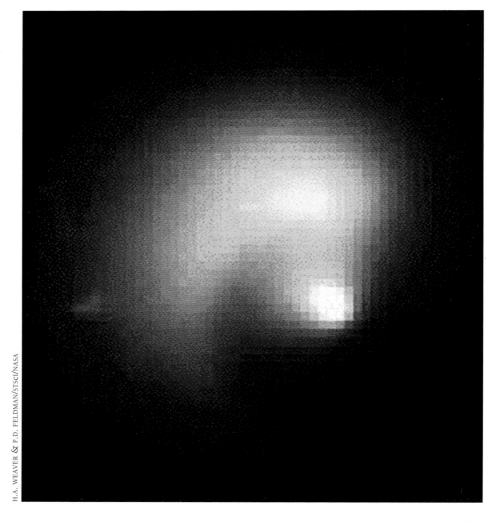

H.A. WEAVER & P.D. FELDMAN/STScI/NASA

but ion tails twist and turn as the solar wind affects them. As with the dust, Hale-Bopp has been quite productive of gas, so the ion tail, too, should be spectacular.

To explore details in the comet's head (the coma) calls for a telescope. At lowest power, the coma will be a soft glow with a brighter core. You may see one or two luminous shells of gas within the coma. These come and go as clouds of gas erupt from the nucleus and are reshaped by the solar wind. Often they change from night to night or over several hours. At high magnifications, examine closely the bright core of the coma. This is called the false- or pseudo-nucleus. It looks like a real nucleus, but it's just the brightest part of the coma — the real nucleus is much too small to be seen from Earth. Another feature to look for are geysers of dusty gas that are mushrooming as they merge with the coma. Sometimes these look like spikes, other times more like fans of light. Irresistably, they suggest water squirting from a spinning sprinkler. While you can't see the eruption sites on the real nucleus, careful observation of changes in the jets hints at the unseen nucleus' rotation.

Measurements of Comet Hale-Bopp's outbursts in the summer of 1996 gave a rough rotation period of 19.5 days, but this needs further study.

High-power observing, whether of jets, shells, or any other feature, calls for a settled and tranquil atmosphere, something not always available. The problem is like looking across a parking lot on a summer day. Warm air burbles up to distort the "seeing," as astronomers call it. The same thing happens in celestial observing. The remedy is to find a location with few heat sources in the line of sight to the comet. Parks, school yards, and playing fields are usually good. Looking out an open window or across roofs, chimneys, or parking lots are poor choices. You won't always have an ideal situation, and when bad seeing occurs, the best course is simply to reduce the magnification until the image stops trembling too much to look at.

Now let's take a detailed look at where and when to view the comet — and what you'll likely see.

COMET-WATCHER'S LOGBOOK

What follows is an observing calendar that runs from January 1997 past the end of 1998. It describes where the comet is in the sky, what it's doing, and what you should look for with it. The description also profiles (where there's a difference) what observers in the northern and southern hemispheres will see. A word on distances: planetary scientists use the astronomical unit (AU) as a handy yardstick within the solar system. One AU is equal to Earth's average distance from the Sun, 93 million miles or 150 million kilometers. For comparison, Mars orbits at 1.52 AU, Jupiter at 5.20 AU, and Pluto at 39.44 AU.

January 1 to 15, 1997

Set your alarm clock for early morning! Horrible as it sounds, during January and February the best time to see the comet is the chilly hours before morning twilight. As January opens, the comet lies among the faint stars of Serpens Cauda (the Serpent's Tail), amidst the Milky Way. But it is moving toward Aquila and growing brighter and more prominent each morning. The comet will resemble a small hazy patch, with a brightness of about 3rd magnitude. It may be hard to locate at first, especially from an urban or suburban location. From a location with a low, unobstructed horizon (a schoolyard, ball field, park), use binoculars to scan for it. Once you find the fuzzball, try to see it with unaided eyesight.

Around 6 a.m. on January 1st, look for the comet about 35° to the lower right of 1st-magnitude Vega in Lyra (the Lyre), and about the same distance to the left of −4 magnitude Venus. Both comet and Venus will be quite low in the sky. A week later, the predawn sky on January 6th will be especially beautiful, with an "old" Moon lying low in the southeast, close to Venus. Hale-Bopp will stand to the left of them and higher, with its tail becoming quite noticeable. Over the next couple of days, Mercury will rise to meet Venus as the comet continues its day-to-day motion toward the 1st-magnitude star Altair in Aquila.

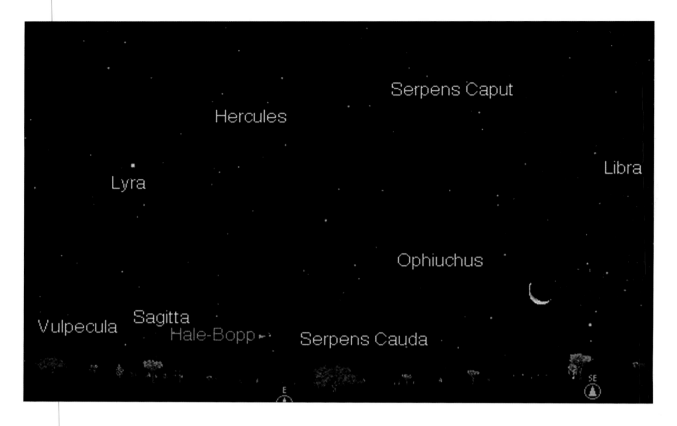

January 6, 1997: Moon and comet low in the east before sunrise (northern hemisphere). The comet is to the left of Serpens Cauda.

With a telescope, try to locate shells or layers of gas in the coma, and recheck them on succeeding mornings. Also, estimate the length of the tail. Now's the perfect time to open a comet log to record your observations. It doesn't have to be fancy — a school notebook works fine. For each entry, record the date, time, weather conditions, the instrument you are using (if any), and whatever you see.

(**Southern hemisphere observers**: skip ahead to April 19th.)

On January 5th, Hale-Bopp lies 2.50 AU from Earth and 1.70 AU from the Sun. It is still beyond the orbit of Mars.

January 16 to 31, 1997

The comet continues to brighten. Between January 15th and 31st, it crosses Aquila, finishing the month just 7° above Altair — about as wide as three fingers held at arm's length. The comet's gas tail will now be a luminous streak pointing toward the upper left, directly away from the Sun, which lies below the eastern horizon. The Moon is full on the 23rd and Last Quarter comes on the 31st; this means moonlight will lighten the sky and reduce the apparent length of the comet's tail during the latter half of this observing period.

Comet Hale-Bopp is now about as far from the Sun as Mars orbits, although the comet is approaching the Sun from high above (that is, on the northern side of) the ecliptic, the plane in which the planets circle the Sun. Examine the comet for activity in the coma — shells or jets — and measure the tail. As the comet approaches it becomes more active, plus it lies nearer to Earth so changes are easier to see. The comet's appearance will alter on a week-to-week basis, so keep your observing log up to date. If you have a camera but haven't yet attempted to photograph the comet, now's a good time to start. See Chapter 4 for details.

On January 20th, Comet Hale-Bopp lies 2.23 AU from Earth and 1.51 AU from the Sun, just inside the distance at which Mars orbits.

February 1 to 10, 1997

As February begins, alert observers will notice that the Sun rises earlier each day. If you go outside at 6 a.m. sharp each morning, you'll see that sky has a bit of blue in it that was lacking a month ago. Since late December, dawn has started to arrive sooner every day, as the Sun makes its slow trek north toward the March equinox and the June solstice. On February 4th and 5th, the waning crescent Moon will lie about 45° to the right of the comet, eminently suitable for a wide-angle camera shot.

On February 4th, the comet lies 1.94 AU from Earth and 1.33 AU from the Sun.

February 11 to 20, 1997

Comet Hale-Bopp now lies fully within the Summer Triangle. Predictions are always risky with comets, but Hale-Bopp should be 1st magnitude by now and display a superb tail. It will be the most eye-catching thing to see in the eastern sky before daybreak.

February 11, 1997: Hale-Bopp in the Summer Triangle — looking east before dawn (northern hemisphere).

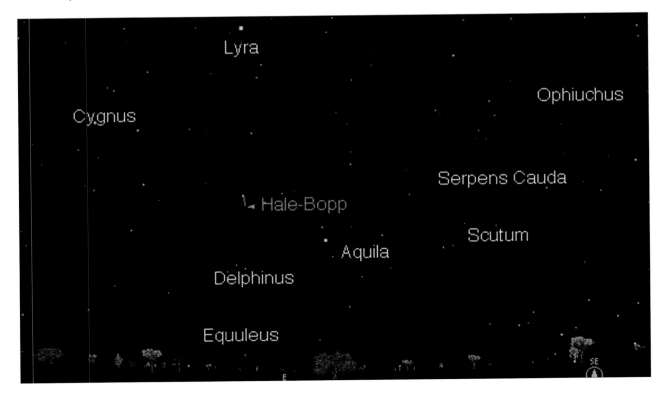

This is a good time to use a telescope to try exploring the comet's head and false-nucleus for activity. The comet has not experienced this much solar heat for more than 4,000 years. Its ices are boiling off rapidly, liberating loads of dusty gas. These stream away from the nucleus into the coma and thence into the dust and gas tails. If it seems hard to capture the comet's appearance in words, try a small sketch. It doesn't have to look professional, just show what you're seeing. Use a soft pencil (#2 or HB grade or softer) and smudge the marks to smooth them. It will look a little odd at first, but such "negative" images are preferred by astronomers because they are easier to do accurately.

Seeing the comet's tail at its best means using a dark sky, and that means setting your alarm a little earlier. Plan to be outside and looking east by 5:30 a.m. at the latest. *On February 14th, Comet Hale-Bopp lies 1.75 AU from Earth and 1.22 from the Sun.*

February 21 to 28, 1997

During the last week of February, the comet lies in Cygnus (the Swan), fully embedded within the Milky Way. It may seem strange to view the stars of a summer's night across snow-covered fields or roofs, but in half a year's turn of the seasons they will make their evening appearance at the appointed time. The comet meanwhile must contend with the growing moonlight during this viewing period — Full Moon comes on February 22nd. But Hale-Bopp is only a month from its closest approach to Earth and the Sun, so it is daily becoming more interesting and bright.

On February 24th, the comet will be 1.57 AU from Earth and 1.11 AU from the Sun.

March 1 to 7, 1997

As the calendar flips over into March, the comet is poised to begin its grandstand show. Hale-Bopp remains a striking sight in the east-northeast before sunrise. In the first week of March, the Moon passes Last Quarter (on the 2nd) and then drops down each morning a little closer to the eastern horizon, vanishing after the 5th. Comet Hale-Bopp lies near Cygnus, about 15° to the lower left of the 1st-magnitude star Deneb. The comet is moving day by day to the northeast on the sky, which means down toward the horizon as you look east in the last hour or so before dawn. The constellation it is moving toward is Lacerta (the Lizard), a dim zig-zag of 4th-magnitude stars that's hard to make out except under a dark sky. The head of the comet will be visible as a fuzzy patch with a bright starlike core, and you'll see a strong gas tail pointing toward the upper left.

On March 6th, the comet will be 1.42 AU from Earth and 1.03 AU from the Sun. About three weeks to go until perihelion — closest approach to the Sun.

March 8 to 14, 1997

All right! One final week of the Dawn Patrol and then you can shift comet-watch operations back to the evenings. Comet Hale-Bopp is now moving noticeably each day (gauge the movement by measuring its distance from nearby stars). The comet

The comet's tail stands generally upright in the west, but compare its angle to the horizon with what it was two weeks ago and you'll see a change. What happening is that the Sun is moving north toward the June solstice point. And the comet's bluish gas tail, blown like a banner in the wind, points directly away from the Sun. As the Sun moves, so must the comet's tail.

Use your telescope to check on coma activity. What are the jets doing? Can you guess at a rotation period for the nucleus? How many shells populate the coma? What about structure — streams, knots — in the gas tail? The growing moonlight (First Quarter comes on the 18th) cuts into the comet's visibility, but the view is hardly diminished.

On April 15th, the comet will be 1.51 AU from Earth and 0.95 AU from the Sun.

April 19 to 25, 1997

For northern hemisphere viewers, Comet Hale-Bopp continues its day-by-day shift to the left toward Orion. At the end of this observing period, the comet stands about 9° above the Pleiades. This fits neatly into the field caught by a telephoto lens, and will make a spectacular photograph, especially if several clear nights in a row let you shoot a series. Observationally, the comet has much to offer as it continues its post-perihelion activity. Look for jets and eruptions in the coma, and be alert for disconnection events and knots in the gas tail.

Southern hemisphere viewers can start to catch sight of Hale-Bopp low in the northwest as evening twilight fades into night. The comet lies below Taurus and Orion, the latter being easier to see as it is higher. Now's the time to open your comet logbook and start recording what you see.

On April 20th, the comet will be 1.59 AU from Earth and 0.97 AU from the Sun.

April 26 to May 2, 1997

Seen from the northern hemisphere, the comet's place in the sky tells you it will soon slip from view. Each evening at the same time it lies noticeably lower in the twilight and its tail is shorter. Get out every clear evening and record whatever you can — by eye, camera, or sketchpad. You won't have much longer to do so.

In the southern hemisphere, Comet Hale-Bopp's visibility grows better, although it remains low in the northwest after sunset. Despite the poor seeing that comes with low-elevation viewing, use a telescope to examine the head of the comet. Look for jets, fans, and haloes of light that signify outbursts of dusty gas from the comet's icy nucleus.

On April 30th, the comet will be 1.75 AU from Earth and 1.05 AU from the Sun.

May 3 to 16, 1997

This period marks the last in which northern hemisphere observers can get a good look at the comet. Hale-Bopp's grand apparition is winding up, nearly two years after it began. On May 5th, the comet crosses the ecliptic southbound. The event is

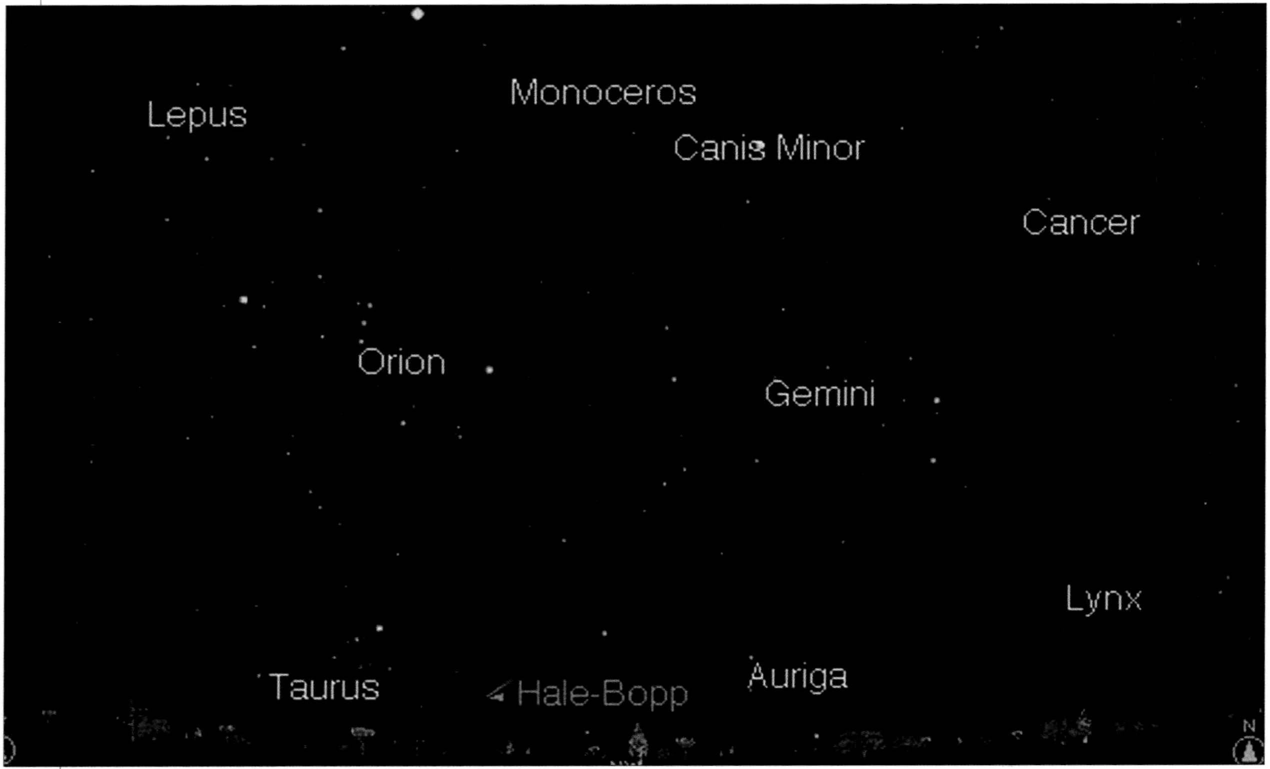

Lepus

Monoceros

Canis Minor

Cancer

Orion

Gemini

Lynx

Taurus   ◄ Hale-Bopp   Auriga

N

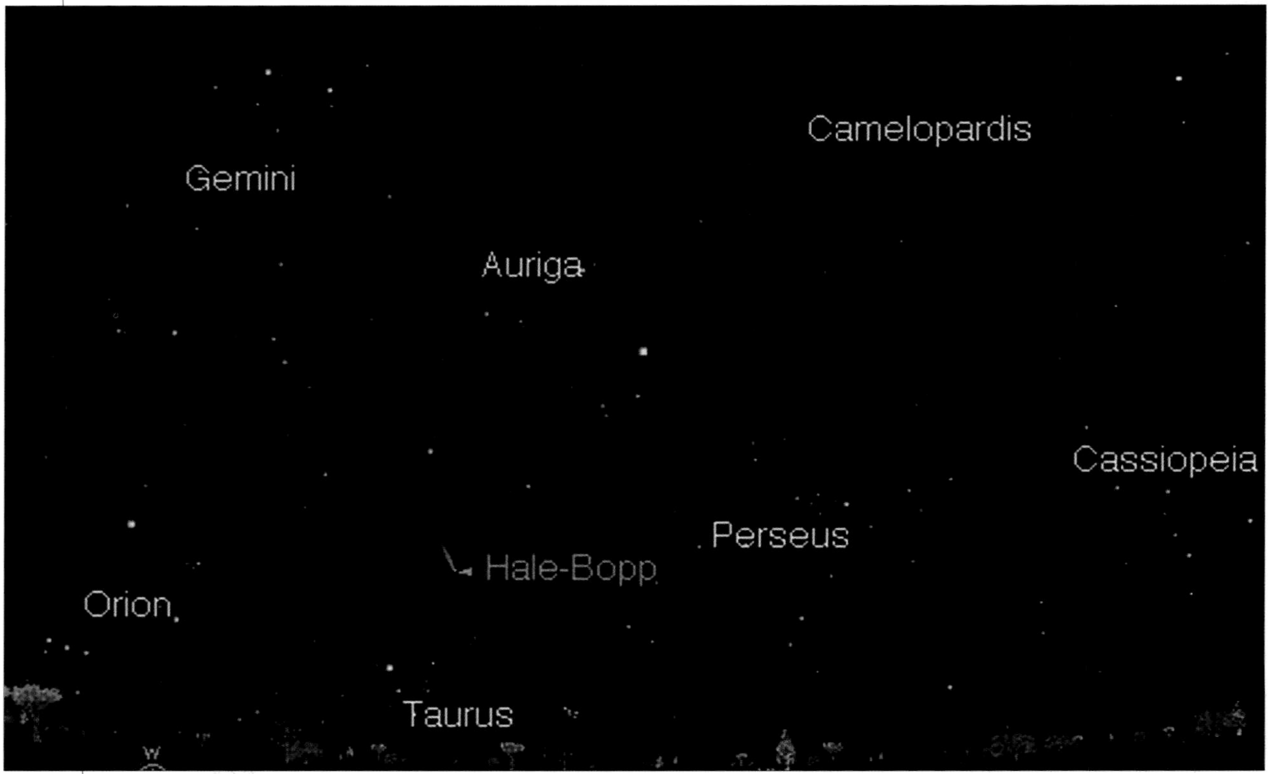

Camelopardis

Gemini

Auriga

Cassiopeia

Perseus

Hale-Bopp

Orion

Taurus

W

more symbolic than practical, but it underscores the comet's imminent departure. In the evening on May 8th, a slim crescent Moon passes just south of the comet to make a fine pair for viewing with low-power instruments — binoculars would be ideal. And with the comet itself, take advantage of every viewing and photographing opportunity the weather allows.

From south of the equator, the view is better. Because of the dateline, observers in Australia and New Zealand see the conjunction of the Moon with the comet in the evening on the 9th. The growing moonlight (First Quarter Moon comes on the 14th) will interfere with the comet a little in this viewing period, but it will still be quite bright and its tail a distinct sight.

On May 10th, the comet will be 1.92 AU from Earth and 1.14 AU from the Sun.

May 17 to 31, 1997

Sometime soon, northern hemisphere viewers will lose sight of Comet Hale-Bopp. Orion and the comet are disappearing into the evening twilight; sunsets occur later each night, helping to chase this part of the sky off-stage. Any clear evenings you get, take a *really* good look. This will be your last view of the comet except for a brief opportunity in the autumn when Hale-Bopp will be much fainter. (Now skip ahead to the entry for August and September.)

For southern observers the apparition continues full force, with the comet placed in the northwest every evening. It is moving toward the reddish star Betelgeuse in Orion. Although the comet is low in the twilight, it will be fairly bright. A series of photos, taken over a week of evenings will clearly show the comet's movement near Orion. With a telescope, note for the logbook the changing coma, with its luminous spikes, jets, fans. and shells of dusty gas.

On May 25th, the comet will be 2.17 AU from Earth and 1.31 AU from the Sun.

June 1997

Southern hemisphere viewers can keep Comet Hale-Bopp in sight during all of June, but the first two weeks are better than the last two. The comet is brighter and slightly higher in the evening sky. As June begins, it lies near the bright orange star Betelgeuse in Orion, but the comet's course takes past the star as the evenings roll by. On June 7th, a crescent Moon stands to the comet's right. The bright planet Venus lies below the Moon, but it will be hard to discern without very clear skies and a flat northwestern horizon.

After passing Betelgeuse, the comet continues toward the southeast on the sky. This path is taking it into a star-poor region of sky, the constellation Monoceros (the Unicorn). By this time the comet's brightness will have noticeably diminished. The tails will have shrunk, and coma activity should be diminishing. Yet this is also when comets can surprise observers with sudden outbursts, so it'll repay you to keep a close watch on the comet as long as you can. Full Moon on June 20th makes the comet more difficult to view. By the end of June, Venus will be higher each

May 1, 1997: After sunset, look below Orion in the west to spot Hale-Bopp (southern hemisphere).

May 2, 1997: Departure draws nigh — Hale-Bopp is getting low in the west after sunset (northern hemisphere).

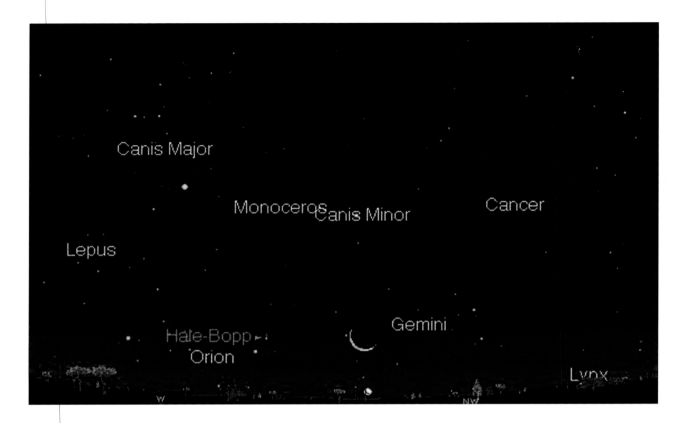

June 7, 1997: Moon and comet pose together in the evening twilight (southern hemisphere). Venus is on the horizon below the Moon.

evening than the comet, and Orion will be gone. The easiest signpost to use for locating Hale-Bopp will be the star Sirius, the brightest in the entire sky. The comet lies about 15° to the lower right of Sirius — a little more than a palm's width held at arm's length.

On June 14th, the comet will be 2.44 AU from Earth and 1.55 AU from the Sun.

July 1997

As July opens, Comet Hale-Bopp is a hard-to-find object low in the western sky in the evening twilight. Try using Sirius (which is also going to disappear soon) as a guidepost to the comet. Look about 13° — one palm-width — the lower right of Sirius, and you'll probably need binoculars to be sure of seeing it. On the 6th, a crescent Moon appears to the comet's right, with Venus to the Moon's upper right and elusive Mercury below the Moon. (View on a very clear evening from somewhere with flat western horizon.)

However, after the beginning of July, the best views of the comet will be in the eastern sky before morning twilight is well advanced. During the month, the comet's southeastward course keeps it in the vicinity of Sirius. On the 10th Hale-Bopp lies 12° to Sirius' lower left; by the 31st, it's 9° directly below the star. Also on the 31st, the waning crescent Moon lies well to the comet's left, but makes a photogenic group with Taurus, Orion, Sirius, and the comet.

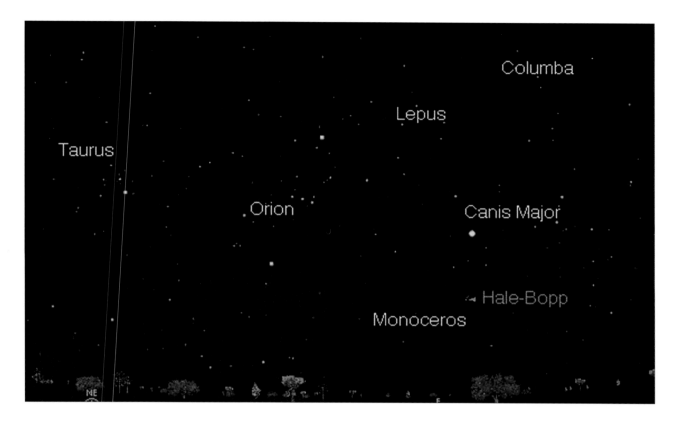

On July 14th, the comet will be 2.75 AU from Earth and 1.93 AU from the Sun. The comet lies at a distance from the Sun about equal to the inner edge of the main asteroid belt, but the comet is already nearly 1 AU below (south of) the ecliptic, the plane the planets and asteroids orbit within.

July 26, 1997: The Dawn Patrol — look for Hale-Bopp before sunrise in the east, below Canis Major (southern hemisphere).

August to September 1997

With August on the calendar, look for Hale-Bopp in the predawn sky. The comet continues to move southeast relative to the background stars, but its motion has greatly slowed. As a result it hangs in the vicinity of Canis Major (the Greater Dog) and Puppis (the Stern) until October. During this time, the comet's retreat from the Sun — plus its increasing distance from Earth — will diminish its brightness to the point where it is a telescopic object only. Its track takes it near one naked-eye star: between August 31st and September 2, the comet will pass close to the 3rd-magnitude star Xi (ξ) Puppis, which will make for interesting viewing (and photography). But that's just the prelude. By the end of September, the comet is approaching the starfields of northern Vela (the Sails), where it will glide past many stars in this increasingly rich region of the southern Milky Way.

Northern hemisphere viewers have a final opportunity to catch the comet late in September. Viewing is terrible: the comet lies very low in the south-southeast as morning twilight rapidly brightens the sky and makes the comet hard to see.

Around September 25th, look for the comet with binoculars about 26° to the lower left of Sirius, the brightest star in the sky. Light from a waning Moon adds difficulties, but the real problem is low elevation and morning twilight.

On August 28th, the comet will be 2.99 AU from Earth and 2.49 AU from the Sun.

October to December 1997, and after

For southern observers, Comet Hale-Bopp is well-placed for viewing — if you don't mind keeping early hours. Among the stars of Puppis, the 6th-magnitude comet is an easy sight in binoculars and small telescopes. From October 5th to 7th the comet passes close to the 2nd-magnitude star Zeta (ζ) Puppis, a pretty conjunction to watch, and on October 24th, it will pass even closer to 2nd-magnitude Gamma (γ) Velorum. After these encounters, the comet's path lies first southward, and then it turns west by December. On New Year's Eve, Comet Hale-Bopp will lie about 12° south of Canopus, the second-brightest star in the sky and the main luminary in Carina (the Keel). Observers should follow the comet as long as they can, noting the comet's behavior and especially looking for changes in the coma. By now this will probably be pretty featureless, except for a brighter core — but comets are always unpredictable.

In January 1998, Comet Hale-Bopp (a faint object for medium or larger telescopes) passes near the Large Magellanic Cloud, which should make for interesting photographs. It then moves in a large loop through Dorado (the Goldfish) and

October 10, 1997: One last look — the comet is very hard to find low in the southeast in the predawn sky (northern hemisphere).

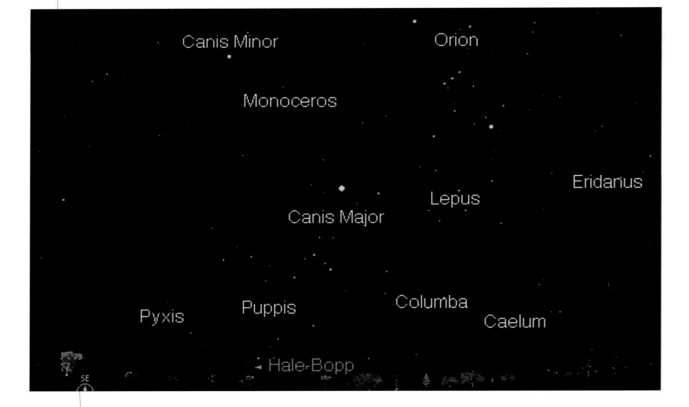

Pictor (the Painter's Easel), passing near the star Canopus in July and August. Finally, it ends 1998 in the dim constellation of Mensa, and the comet will remain in this part of the sky well into the new century. Exactly when the comet disappears from view will hinge on what instrument the observer uses. But most backyard astronomers will lose sight of the comet forever sometime in 1998.

How late into October 1997 northern viewers can follow the comet will depend greatly on their latitude. At mid-northern latitudes, the comet will disappear sometime in the first half of October. But from latitudes closer to the equator, the comet may linger well into November. During October and November the comet's track on the sky lies almost due south, the worst possible course. The comet is also fading steadily (barring eruptions and outbursts). It isn't in the nature of Nature to end comet apparitions with a dramatic flourish, so the comet's diminuendo finale is completely in keeping with its arrival, two and a half years ago.

On October 17th, the comet will be 3.17 AU from Earth and 3.08 AU from the Sun.

The departure of a comet that has put on as great a show as Hale-Bopp inevitably leaves a void for both professional astronomers and amateurs. But it's worth remembering that Hale-Bopp is not the last bright comet you'll likely see. New comets are discovered at a rate of about one a month. Most, it's true, are doomed to remain faint and telescopic. But there's always the chance that the very next comet to be discovered will outshine any previous one — including Comet Hale-Bopp.

CHAPTER 4 — *Shooting a Comet in the Sky*

JOSHUA VAUGHN

Comet photography doesn't have to be complicated. Joshua Vaughn shot Comet Hyakutake against the red rocks of Sedona, Arizona, by putting his camera on a tripod and exposing for several minutes. Every location offers some way to inject a bit of drama into a photo — look around and use your imagination.

With fast film, a tripod, and a camera that permits time exposures, you can take a striking set of images recording your experiences with the comet. You can display them as artwork or show them off to friends — and you can also submit them to astronomy publications and perhaps see them in print. Sound interesting? The first step in the procedure is to decide the film to use. We'll examine what's most suitable for comet photography. Then we'll identify the key features to look for in cameras and lenses. Next, we'll take a look at what kinds of comet photos you can take (and what they show), and finally we'll outline some steps to ensure best results when the images are processed commercially.

The right film. Choosing a film depends partly on taste, partly on what you want from your comet photos, and partly on where you photograph from. To sort these issues out, let's start with the most basic question — black & white or color? Some photographers may disagree, but I recommend you stick with color unless you own a darkroom. Color film is available in more variety and more widely than black & white. Moreover, comets display subtly-colored features. And finally, it's generally easier to take impressive color photos than black & white ones.

The next decision — slides or prints? — is less clear-cut. Slide film offers many advantages. Slides usually render images that are sharp and contrasty; you might think of them as "true-to-life, plus a bit more." And when seen projected, their vibrant colors can make you gasp. Moreover, publications such as ASTRONOMY and *Sky & Telescope* prefer to work with color slides because they reproduce better. Last, in processing there's much less chance of operator error.

But slides have drawbacks. The biggest is that you can't frame one and hang it on the wall. A second drawback is that when used at their highest speed ratings, slide films aren't quite as satisfactory as equivalent-speed print films — they are grainier and their color response is all but impossible to control or adjust. Print films offer more flexibility, plus you can order prints in many different sizes. The main drawback to prints for people who don't have a darkroom is that prints of astronomical objects can easily go awry when made by the typical photo-processor.

Whichever the format, I recommend films of ISO (ASA) 400 or faster. This is the fastest film you can use for photographing the comet from an urban or suburban location, where the sky is heavily light-polluted and bright. However if you can get well away from the city — and I urge this, both for photography and viewing — then use film speeds up to ISO 1600, the current top speed available. (Higher speeds are possible with special processing.) High film speeds deliver a longer comet tail and fainter stars per exposure. This translates into better pictures in less time.

GUS JOHNSON

Gus Johnson nicely caught Hyakutake on April 17, 1996, using a piggybacked camera with a 135mm lens, Kodak 400 Gold print film, and a 2-minute exposure at f/2.8. He tracked the comet by moving his telescope by hand in elevation and direction, a rough-and-ready technique that worked here because he kept the exposure short.

Now for some specific recommendations. These films fall outside what most drugstores and discounters carry in stock, so plan to visit a camera store. The price may be a bit higher, but rolls will be fresher and they will have been stored properly. For slides, go with Kodak's Ektachrome P1600 as a first choice and Ektachrome Elite 400 as a second. Fuji's slide films get mixed reviews from sky-shooters; some love them, but others won't touch them. Ektachrome earns a solid rating with everyone, so it makes a logical pick. For print films, on the other hand, the laurels go generally to Fuji. Try Fuji's SuperG 800 Plus as a first choice and SuperG 400 Plus as the fallback. Fuji's reign at the top may be ending, however. A new print film from Kodak called Pro 400 (or PPF) is creating a stir among astrophotographers. They like its snappy contrast, good response to low light levels, strong color rendition (even when underexposed), and crisp detail.

Some advanced sky-shooters use a technique called hypersensitizing (or just "hypering"). It wrings every bit of speed out of film by baking it at elevated pressures in a special gas. While the results can be spectacular, hypering requires much experience to use correctly and the arrival of a major comet is a poor time to start learning.

Whatever film you choose, shoot one or more test rolls before the comet is at its best. Do this to find the best exposure combination with your specific camera and lens — the camera's light meter won't work reliably (or at all) on subjects like a comet. Set up as you would for a scenic view of the sky (see below). Then take exposures that start at 10 seconds and roughly double with each successive shot. Ten seconds, then 30 seconds, 1 minute, 2 minutes, 5 minutes, 10 minutes, and so on. Continue the series as long as you like, but keep a logbook so you can determine what worked.

Over page. Simple equipment plus painstaking care netted Chuck Vaughn this magnificent image of Comet Hyakutake on March 24, 1996. He piggybacked his camera (with a 350mm lens) on a telescope, tracked the comet, and took two 10-minute exposures at f/2.8 on Fuji SuperG 400 Plus print film. He sandwiched the two negatives to make the final print.

Once you find a film you like, stick to it. Don't get bogged down in trying new films in the middle of the comet's apparition. You'll have a much higher probability of snaring excellent results if you're not continually experimenting as you go for the gold.

The right camera and lens. Choosing film was easy. The situation with cameras is more complicated. Most new cameras are loaded with automatic features to make ordinary picture-taking — well, a snap. But features that benefit the general photographer become drawbacks when you turn to sky shooting. To a large extent, photographing the sky with a modern all-automatic camera is an exercise in deliberately thwarting built-in safeguards. From the camera's point of view, the typical sky scene is badly lit and has no central object on which to focus. So the camera tries to prevent you from taking the picture in the first place and when you insist, its reponse is to turn on the flash. This is totally useless, except perhaps to light up a foreground that's better kept dark for pictorial reasons. And wait — worse can happen. The automatic exposure control may not know when to close the shutter on its own and you'll probably need to override the lens aperture gizmo and set the focus to infinity by hand. Otherwise, the camera may end up choosing a focal point of its own somewhere between the rhododenrons and Alpha Centauri.

To determine if your camera is able to take comet photos, locate your owner's manual. See what it says about taking time exposures at a focus and an aperture you select by hand. If it has clear, intelligible instructions, you're all set. But if it doesn't, or if you have a point-and-shoot camera, I recommend you do what experienced sky photographers do. Buy an older camera that lacks whizbang features, but places all the controls under manual command. Check what a camera store has in used equipment. Look for manual focus and aperture controls.

Bright turquoise seen by many in the head of Comet Hyakutake was no illusion. It came from ionized carbon monoxide and carbon molecules fluorescing in sunlight. David Churchill used a 200mm lens, hypersensitized Fuji SuperG 800 Plus print film, and a 30-minute exposure at f/2.8. ("Hypering" is an advanced technique that increases a film's receptivity to light.)

Dedicated sky photographers buy or devise an equatorial platform to let them keep one or more cameras aimed on the stars automatically. Such sky-trackers can often be adapted from commercial equatorial telescope mountings. (The comet is Hyakutake, seen from southern Arizona in March 1996, on a night when it was near Polaris — note the Little Dipper curving up to the right of the comet.)

Fields of View

28mm lens	46° x 66°	135mm lens	10° x 15°
35mm lens	38° x 54°	250mm lens	5.5° x 8.2°
50mm lens	27° x 40°	500mm lens	2.7° x 4.1°
70mm lens	19° x 29°	1000mm lens	1.4° x 2.1°
100mm lens	14° x 20°		

Thumbnail Guide to Comet Photos

28mm to 70mm lenses	scenic views
70mm to 1000mm	comet + stars imaging
1000mm or greater	close-ups on head and nucleus

Shutter speeds don't have to be accurate; what counts is having a shutter that opens and closes reliably. Simple manual cameras aren't fashionable at present, which is good news because you can often pick up a highly serviceable "junker" camera for under $100, often *way* under.

Such cameras work fine for simple views of the comet. But if you intend taking photos through a telescope, or if you want to image the comet at different scales, the ability to remove one lens and substitute another is paramount. You'll also need to be able to view through the lens for focusing. In a used camera, the optimal choice is a 10 to 20 year old basic model SLR (single-lens reflex). You can also get a new Pentax K1000 camera body for about $150 from mail-order discounters. Check the ads in *Popular Photography* or *Shutterbug* magazine.

At the other end of the technology spectrum is electronic imaging. Using smaller versions of the sensors on big professional telescopes, some amateurs have started doing the kinds of imaging that can dissect, analyze, and process the daylights out of a comet's picture. The results can be amazing (if not exactly pretty), but as with hypersensitizing, the prerequisites in experience (and money) tend to put this out of the reach of beginners.

With lenses, one having a lower f/number is better because it lets more light through. What about zooms? Experienced astrophotographers shun them and for good reason. Many do not focus as crisply as do single-focal-length lenses, and in a photographic sense they are "slower," letting less light through. What focal length lens is best? That depends on what you wish to photograph. Every lens captures a different scene according to its focal length and the size of the film frame. This guide to the fields of view assumes the negative measures 24mm x 36mm, which is the standard size for full-frame 35mm cameras.

Kinds of photos. Planning photography of Hale-Bopp starts with understanding that comet photos fall into three general categories. Each of these calls for different equipment and techniques.

The first kind of photo shows the comet about the way it looks in the sky, perhaps with a foreground of trees and houses. The second kind captures the comet at closer range, showing it in space with the head and tail displayed against a starry backdrop. (These images are often the most dramatic and beautiful.) And the third kind zooms in on the comet's head to catch the coma in detail. These photos can reveal bright swirling jets and shells of luminous gas coming from active areas on the nucleus. Astronomers and advanced amateurs usually take this kind of photo (in addition to the others), because they bring you closer to the comet's heart.

So what's the best kind of photo to take? There really isn't any "best" photo because each of these captures some part of what you see when Hale-Bopp (or any comet) comes by. Let's now look at the kinds of photos in more detail.

THE "SCENIC VIEW" PHOTO

Photos that capture the comet in the sky over a silhouetted foreground come closest to capturing the naked-eye scene. It's also the easiest comet photo to take. All you need is a camera, a tripod, some fast film, and a site that affords a good view. Getting away from city lights is a big plus, as is shooting on a moonless night — both provide darker skies. While it isn't always possible, the advantages of a trip to dark skies are great and well repay a little effort on your part to juggle work and family commitments. Do what you can.

Let's assume that you have a camera you can control (see above), it's loaded with high-speed film (ISO 400 or faster), you have a tripod, and a dark-sky location to shoot from. Attach the camera to the tripod and set it up facing the comet. If you can frame the view with trees or a handy mesa, so much the better. Attach a cable release if you have one to the shutter button to trip the shutter more smoothly. Open the lens to its lowest f/stop and set the focus to infinity (the ∞ symbol). A flashlight with red tissue over the lens can help you set the controls without ruining night vision.

Take your time setting up. Real comets don't zip across the sky like the ones in TV ads. Nonetheless, comets *are* moving and they have two motions photographers have to pay attention to, one large, the other small. The small motion is that of the comet in space, moving relative to the stars. For camera-on-tripod photography, you can ignore this. The second motion — caused by the sky's daily rotation — is worth considering. Every hour the sky wheels westward 15° due to Earth's rotation. The trick lies in exposing long enough to record the comet and stars, but not so long that they smear too much. How long is too long? The answer lies in how much star trail you're willing to tolerate. This varies by taste, but the table at right gives a few rules of thumb. It assumes you are using a camera with a 50mm focal-length lens and don't enlarge prints beyond 5x7 inches. If you are using a wide-angle lens (28mm say) you can roughly double these times and with a small telephoto lens (85 to 100mm), halve them.

"pinpoint" stars	20 seconds
slight star trailing	1 minute
noticeable star trails	2 minutes and longer

Star trails aren't necessarily bad, and comets are inherently fuzzy objects anyway. But let's say you want stars that look fairly pinpointy, so you'll keep the exposure to 1 minute. How much comet can you record in that time? With ISO (ASA) 400 speed film and a lens set to f/2 or f/2.8, you can record the star-like nucleus of the comet, the brightest part of its tail, and the brighter background stars. Foreground objects (unless lit) will appear as black silhouettes. The sky itself will appear either as black, dark grey, or a grayish-yellow-green, depending on how brightly lit it is from city lights. Obviously, the brighter the sky, the less comet you'll record, which is why it's worth some effort to get out of town.

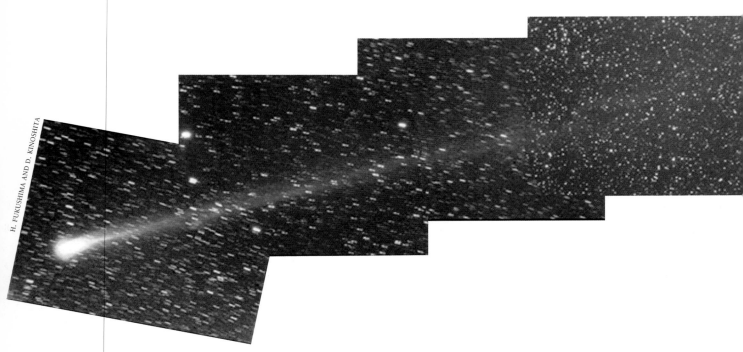

H. FUKUSHIMA AND D. KINOSHITA

Cut-and-paste is the way to handle a l-o-o-o-n-g comet tail. This composite by two Japanese photographers recorded Comet Hyakutake's lengthy tail in four electronic images taken March 25, 1996. Their idea works with film, too. Shoot an overlapping sequence down the tail, make duplicate prints, and get out the scissors and glue.

Now take a longer exposure. How long you can go is something that your test exposures should have shown. As you work through the exposure sequence, remember to use care when winding the film between frames. Cold film becomes brittle and can snap. Also, in dry air, rapid winding often creates little bolts of static electric lightning within the camera. These will ruin any exposure they strike. And watch out for dew or frost on the lens. Use a lens hood to slow its formation, and if necessary put the lens into a warm pocket for a while between exposures.

Give a thought or two to esthetics. Mentally, your focus lies on the comet itself, but think of the people you'll show this to. A little drama isn't out of place. Try to find a site where you can frame the comet in a notch in the hills, or between a few trees, or over the flat expanse of a lake. This is a chance to give artistic sensibilities some scope.

And once the comet has gone, don't let the sky-shooting outfit rest idle. It is outstandingly well-adapted for photographing auroras, meteor showers, close groupings of the Moon and bright stars or planets, lunar eclipses, and other celestial events. Use your newfound techniques to capture these as they happen, and make it a habit to check the monthly sky columns in the astronomy magazines for additional likely subjects. You'll soon discover the sky is full of them.

THE "COMET & STARS" PHOTO

Scenic views from a tripod-mounted camera are highly striking, but for a real showstopper, piggyback the camera on a telescope that can track the stars. The telescope must have an equatorial mounting and a clock or motor drive. Follow the manu-

BRAD WALLIS & ROBERT PROVIN

Comet photos don't have to be splashy to be beautiful. Brad Wallis and Robert Provin's image of Hale-Bopp, taken September 15, 1996, shows the growing comet's asymmetric coma and soft colors. They used a 6-inch Starfire refractor as their camera lens and took a 20 minute exposure at f/7.1 on hypersensitized Fuji SuperG 800 Plus print film.

facturer's directions to align it on the celestial pole. With that chore taken care of, the telescope will function as a platform to let the camera stare at any piece of sky for as long as you like. Stars will remain pinpoints, even when the shutter is open for half an hour or more.

This kind of photo captures the "emotional comet." If you work carefully, the image will be breathtaking, and the photo will surpass your eyeball view many-fold. The price is that you need the equatorial mounting and you have to pay more attention to technique. Many telescope manufacturers sell an adapter to piggyback a 35mm camera onto the telescope or mounting. Check what's available for your instrument. When you attach the camera, rebalance the scope to allow for the extra weight or the drive may not track correctly. Second, look through the camera's focuser to check that the field of view doesn't include some portion of the telescope itself.

Take exposures starting at about 2 minutes' duration and running up to an hour, unless the brightness of the sky forbids it. Again, wind the film slowly, and carefully guard against dew or frost with a lens hood. (One time-honored remedy to keep lenses frost- or dew-free is to occasionally squirt the air from a hair dryer over the front element of the lens.) Keep an accurate logbook of your exposures, noting times and lenses used, together with film data. Believe me, you won't remember the important details two weeks from now, let alone next year!

THE "INNER COMA" PHOTO

This isn't territory for novices. The photos described above use the camera's own lens as the picture-taking element. But a telescope itself can become the lens, resulting in a greatly magnified view of the comet's head. In fact, it's often similar to the view looking though the telescope's eyepiece. To achieve this, you sacrifice the broad view — no sweeping tail, no gossamer veils against the stars. Instead, you get close-up looks at the comet's icy heart. These are images that aficionados love (and which often bore non-astronomer friends).

Once again you face a step up in price and technical complexity. You will need an adapter that connects the camera body (minus the lens) to the telescope's focuser. See your scope's manufacturer. Also, because the view captured on film is now highly magnified, it will no longer suffice just to let the mounting track the sky. You'll have to guide the telescope yourself on the comet's nucleus, whose motion will differ from that of the stars. This calls for a separate guide scope or guiding eyepiece with cross-hairs, and a means of precisely controlling the telescope's movement in both right ascension and declination.

If the idea is intriguing but you have never done this kind of astrophotography, be prepared to spend some serious money on an excellent mounting and associated paraphernalia. And also plan to spend several intensive weeks learning to control the instrument. The resource guide in the back lists some astrophotography books to help steer you right. Honestly, I don't advise this kind of photography for beginners,

Almost an octopus! Comet Hale-Bopp's active nucleus had spawned seven jets at the time this electronic image was taken by Australian amateur astronomer Gordon Garradd. He carefully processed and false-colored it to reveal the jets. The date was October 7, 1996, when the comet was at a distance from the Sun three times greater than Earth.

GORDON GARRADD

and the comet's apparition is a poor time to start learning the techniques. The schedule is just too unforgivingly short, and the cold weather during the best part of the comet's apparition only increases the difficulty.

OK, WHAT HAVE YOU CAPTURED?

When the slides or prints are processed, sit down with your logbook and make a note on each regarding the date, time, film, and exposure particulars. With prints, write on the back with a fine-point indelible felt tip marker (I recommend Sanford's ultra fine point Sharpie)

Ulysses Comet Watch Program

Dr. John C. Brandt
Laboratory for Atmospheric and Space Physics
Campus Box 392, University of Colorado
Boulder, CO 80309
USA
http://miranda.colorado.edu/ucw/ucw.html

and write very lightly. Ballpoints and roller-balls dent the print and make them ugly and unpublishable. Also put your name and address on each photo. This is a nuisance to do. But if you skip it and submit these for publication anywhere, the publisher will likely reject them no matter how good the pictures are. Magazines get hundreds of submissions every week and haven't the staff time to take care of housekeeping details like this. It's easier simply to reject the photos.

Now let's take a look at the images. All right, some didn't come out. But the ones that did — ! Wow!

If you line up images taken with the same equipment on different dates, you can track the comet's path through the sky. Photos also record its changing appearance, which may have altered more than you were aware at the time. Examine the images closely. Can you discern separate gas and dust tails? The gas tail is made of ionized gas, the other is silicate dust particles. Gas tails usually photograph bluish-white, while dust tails (reflecting sunlight) often look distinctly yellowish. Moreover, a dust tail appears broad and fanlike while the straight-pointing gas tail often has knots and kinks in it, like a flag on a windy day caught in mid-flap. Knots in the gas tail signal turbulence in the solar wind, and you may even see a portion of gas tail apparently snipped off.

Such "disconnection events" are of great intererest to comet and solar scientists because they betray changes in the otherwise-invisible wind of particles fleeing from the Sun. If you photographed a disconnection event and have a print or slide you can part with, write to the Ulysses Comet Watch Program at the University of Colorado. The project uses amateur and professional photos to study the changes in the solar wind by their effects on comet tails.

Comet tails can also alter appearance as our viewing angle on the comet changes. Compare your images with the description in Chapter 3 of where the comet is in the solar system. What relations can you see? If you have photos that zoom in on the nucleus, you'll see changes in jets, fans, and shells, as active regions on the nucleus change. Also, the nucleus is rotating and this sprays the jets much like swinging a garden hose around. Put your best images in a time sequence and see what they reveal.

TIPS FOR BETTER PROCESSING

If you have never processed photographs, learning to do it with your first comet images is a bad idea. There's too much at stake with these pictures. Instead, plan on having them processed by a local photo-finisher. This is especially important if you shoot color print film.

Commercial photo-finishing uses automated equipment to develop, fix, wash, and dry your rolls of film. When the film is dry the machine chops each roll into strips typically 4 frames long (for prints) or single frames (slides). Since the machine counts the frames automatically, you can help it put the cuts in the right places by beginning each roll of film with one or two frames of daylight subjects. These help calibrate the machine's counter. Does the thought of a machine mindlessly chopping your precious comet film make you uneasy? Then do as most astrophotographers do — ask that the film be left uncut.

While slides are simple and cheap to process, many people prefer to shoot print film because the prints are easier to show to others. Unfortunately, commercial photoprint-making has several pitfalls to stymie novice sky-shooters.

The first is incredibly basic: sharp focus. A lot of prints from quickie photo shops are unsharp. Most customers assume the fault is theirs, but it often isn't. Inspect the negatives carefully using a magnifying glass. If the print appears blurry but the negative is sharp, the error lies with the print machine operator, not you. Point out the error and ask for a reprinting. When few customers do this, machine operators become sloppy and careless.

Rough ride. Comet Hyakutake's gas tail displays a lot of turbulence in this composite image made from two blue-senstive black and white photos. Besides being pretty, such images can help astronomers study the solar wind, a flow of charged particles that fills interplanetary space. (Kodak IIaO plates, Schott GG385 filter, 105-cm Schmidt camera, 15-minute exposures.)

KISO OBSERVATORY, UNIVERSITY OF TOKYO

N

E

4.0 deg.

TOM OSYPOWSKI

When Comet Hale-Bopp has come and gone, the record of its visit that you create on film will keep its memory fresh and alive. (Tom Osypowski made this souvenir of Comet Hyakutake on March 25, 1996, with a 50mm lens at f/4 using Kodak Ektar 1000 film and a 6-minute exposure tracking on the sky with an equatorial platform mounting that he manufactures.)

The other big pitfall is bad image quality. The problem is that astronomy images are radically unlike ordinary photos. Some intelligent intervention is necessary, on your part and that of the print machine operator. This is why getting to know your local photo-finisher is highly important. Some discount stores and many malls have a photo shop that offers one-hour processing and print-making on the spot. Find one near where you live and test them with a couple rolls of ordinary photos. You might worry they'll think astronomical pictures are too bizarre to bother with, but the inverse comes closer to the truth. Many machine operators are bored silly by endless picnic and birthday photos. Astronomy images are visually striking and usually arouse their interest. By encouraging this, you can get better processing of your prints.

To the print machine, a typical comet photo is a grossly underexposed negative. Detecting a "thin" frame, the machine is programmed to hold back on density to keep the print from becoming hopelessly dark. Good theory — but wrong result. You need to enlist the cooperation and skill of the machine operator. Use common sense and drop by at a time when the place isn't swamped with customers. Also keep in mind that photo processing is a business, and reprinting frames that go wrong eats into the profits. The key to willing cooperation is to make it clear that you will pay for *all* the prints, even the experiments that fail. This lets the operator stop worrying about what the boss is going to say, and concentrate on making the machine deliver excellent sky photos — which it can do easily. For your part, you need to show what the final image should look like, and the operator will translate that into commands for the machine. It may take two or three tries to get a print that makes your heart sing, but when you see it, you'll know the effort was worthwhile.

One more thing. When you get the formula down right for a print, make a couple extras. (If the operator can give you the proper settings, make note of them too.) Extra prints are well worth a little trouble and expense at this point. You're going to want to share the best images with others and making prints now saves redoing the learning curve on a later occasion. In the same way, duplicate slides are a good investment, also.

Comet photography is a lot of fun, and there's no better way to make lasting mementos of the comet's visit. Moreover, the techniques you learn with Comet Hale-Bopp will stand you in good stead with any other comet — and with lots of other sky-shooting opportunities besides.

How to Discover Your Own Comet

CHAPTER 5

Discovering a comet doesn't demand enormous skill. As Chapter 1 shows, you don't even have to be looking to find one! The experiences of Alan Hale and Thomas Bopp may be a little misleading, however. Those were flukes of good fortune. Most comet discoveries by backyard astronomers come after many months or even years spent patiently sweeping the sky. How nice it would be to think that hours of fruitless searching earn you Brownie Points toward the big moment. But the universe doesn't work like that. There's no magic amount of time you have to struggle through before nature relents and awards the prize. "By the time he had put in as much search time as I had," says Alan Hale ruefully, "Howard Brewington [a fellow amateur comet hunter] had already found two or three comets." You might find a comet on your first night looking — or you may never find one in a lifetime of steady searching. Neither extreme is very likely, however.

There are two camps of comet hunters, professional astronomers and backyard amateurs like yourself. Professional comet hunters typically take photographs — on film, on glass plates, or on electronic sensors — using relatively large telescopes. The comets they discover are generally quite faint, at least at the time of discovery. The professionals are working at magnitudes where no amateur can hope to compete. *Well, that's great for them,* you think. *But faced with this kind of competition, what chance have I got?* More than it appears. For one thing, amateur comet hunters outnumber the pros. Second, amateurs can take their instruments where no professional dares to go: right down into the glow of twilight, either before sunrise or after sunset. More than three-quarters of amateur comet discoveries are made when the comet lies less than 90° away from the Sun. And this turf is all theirs. The Hubble Space Telescope cannot point within 45° of the Sun, and the detectors on most professional telescopes would be saturated by the light of the sky during twilight.

One recent discovery story captures the flavor of backyard comet hunting. Japanese amateur astronomer Yuji Hyakutake is a 46-year-old graphic arts specialist. In 1994 he moved to the small town of Kagoshima in southern Japan solely to search for comets. He chose Kagoshima because it had dark skies remote from city lights. There he spent the next couple of years searching for comets with a pair of 20x150 binoculars. These look something like ordinary binoculars, but they are *huge*. Instead of having lenses one or two inches across, these had lenses six inches in diameter. Instead of magnifying 7 or 8 times, his binoculars magnified 20 times. They were so big that he had to mount them on a tripod like an enormous double-barreled telephoto lens.

Hyakutake searched patiently, scanning the skies at twilight and other times of night. Finally in December 1995, on the day after Christmas, he found his first comet. Comet Hyakutake (1995 Y1) was an object of 11th magnitude in the region of Hydra and Libra. It displayed only a small fuzzy disk and no hint of a tail. At best it never brightened to more than 9th magnitude.

But the Comet Hyakutake of December 1995 was not the same one that captured international attention in the spring of 1996. *That* Comet Hyakutake (1996 B2) was found almost exactly one month after the first. As with Hale and Bopp's comet, its discovery was an accident. It happened while Hyakutake was trying to photograph his December comet. On January 31, he says, "I went back to take photos of the first comet. I looked up at the sky where it should have been. However, that particular spot was filled with clouds. I tried to find an area in the sky that was unobscured." The only clear spot was, by coincidence, just where he had found his December comet. When he looked into the giant binoculars' eyepieces, at first he thought there was something wrong with his December comet because there was this fuzzy thing in the field of view. Then with a jolt Hyakutake realized he was seeing a *second* comet! And it was this second Comet Hyakutake that brought the world outdoors for a few evenings last spring as it fleetingly cruised past the Big Dipper before disappearing into the western twilight.

OK, so how do you look for a comet?

Let's start with the hardware. It's simpler than you think. The equipment most often used by amateur comet hunters is not very sophisticated, at least compared to what the market provides. Many successful observers have used relatively small low-power telescopes. What matters most in choosing an instrument are ease of use and a wide field of view. It also helps if you can erect the instrument permanently or at least set it up quickly, since this makes searching simpler and more convenient. And experience shows that those who observe frequently are most likely to catch a comet.

Telescope or binoculars? Most beginners would unhesitatingly choose the largest possible telescope, but the choice isn't that simple. Searching with both eyes at once gains about 40% in effectiveness over using one eye alone. (Your brain perceives more when it works with two visual signals.) Second, binoculars are easier to handle than most large telescopes and this lets you scan more sky quicker. On the other hand, you do need adequate light grasp, which advocates a telescope. (By the time a comet is bright enough to be discovered in 7x50s, someone else has probably nabbed it already.) Is there a compromise? Some comet hunters would opt for a pair of "astronomical" binoculars (11x70 or 20x80 and preferably larger) as being about ideal. Others prefer a 4-inch to 12-inch telescope (reflector or refractor) operating at around 20x to 50x.

Pick nights when moonlight doesn't interfere and the sky is free of haze or clouds. Many comet hunters systematically scan the western sky after sunset and the eastern sky before sunrise. Aim the scope about 10° above the horizon and sweep parallel to it a distance of about 45° to the left and right of the horizon point where the Sun has set or will rise. Use a magnification of about 20x and make sure your sweeps overlap about 10%. Don't rush, but work quickly and smoothly. For an evening search, begin 30 minutes after sunset; in the morning call a halt 30 minutes before sunrise.

Upon discovery, most comets don't look much like a comet. A new one usually resembles a badly focused ball of fuzz, with maybe a brighter center. Here's a scary thought: they look virtually identical to thousands of deep-sky galaxies, star clusters, and clouds of gas. These false alarms are the bane of comet hunting but they're inescapable. This is where knowing the sky in detail comes in. The most successful comet searchers are those who have virtually memorized the entire sky visible from their site. Having this knowledge mentally on call lets you sweep quickly. You pass by dozens of star clusters and galaxies — only to halt when experience says you've encountered something outside the known pattern. (As Isaac Asimov said, the classic phrase of discovery in science isn't "Eureka!" It's more like "Hey, wait a minute....")

Unfortunately, there's no shortcut to acquiring this knowledge. Canadian amateur astronomer Rolf Meier searched for 200 hours and came up with four comets. But Hale logged over 400 hours at the telescope before finding Comet Hale-Bopp. And Californian Don Machholz spent 1,700 hours before he nabbed his first — perhaps a record — yet he now has his name on nine comets in all.

So you think you've found one — what now?

"Verify it!" says astronomer Daniel Green of the Central Bureau for Astronomical Telegrams, the world center where comet discoveries are reported. He says that for every genuine new comet discovery, the Central Bureau sees about five false alarms. Besides already known comets, these usually consist of ordinary deep-sky objects such as galaxies or globular star clusters. Typically, he says, observers caught up in the excitement of a possible discovery fail to really make sure before firing off a report. So here's his checklist to go through with any suspected new comet discovery.

Does it move? Remember how both Hale and Bopp waited to see if the object had moved? This can be nerve-wracking, especially if you are with friends and they are going bananas, but do it. Sketch its position, and then while you wait, thoroughly examine the region of sky in several star atlases. Green cautions against relying on just *Sky Atlas 2000.0* or the older *Atlas of the Heavens 1950.0*. These don't go faint enough — use *Uranometria 2000.0*, for instance. And keep in mind that even a good atlas may not plot every galaxy or fuzzball your telescope shows.

Make sure the image is real. Tap the scope to see if it moves with the other stars and isn't just a reflection in your optics of a bright star nearby. On a photograph, be careful that an emulsion defect or a ghost image isn't producing a "Comet Kodak." To test, observe the suspected region by eye at the earliest opportunity and rephotograph it after you shift the telescope slightly away from its former aiming point.

Get a good position for the object. Chart overlays let you estimate a position to 0.1 minute of right ascension and 1 arc minute in declination. Make enough good position measurements that you can estimate the object's amount of motion and

Comet hunting isn't easy. Most "comets" discovered by eager beginners are false alarms, usually a globular star cluster or a galaxy. Seasoned comet hunters learn to stifle their impatience and triple-check any possible finds. (Brad Wallis and Robert Provin photographed Hale-Bopp on June 15, 1996. They used Fuji SuperG 800 Plus print film and a 30 minute exposure at f/7.1 using 6-inch Starfire refractor as their camera lens.)

PETER CERAVOLO

direction. Note the equinox of the atlas you've used. And note the date and time of your observation in Universal Time (see glossary).

Describe it. What does it look like? Estimate its magnitude (compare with an out-of-focus star of a known brightness) and how diffuse it appears. Does it have a star-like nucleus? Is it lopsided or does it have a visible tail?

Identify yourself and your equipment. Give your name, address, phone and fax numbers, e-mail. What size and type of telescope did you observe with? Magnification(s)? Where were you observing from? If it's a photo, what film did you use? How long was the exposure, at what f/ratio?

One last caution before you send the report. Wait a while and recheck that the object really is moving. If you can't say for certain that it has moved in some direction at a given speed, wait to recheck it the next night. It's much better to be a co-discoverer than blow your credibility as an observer. Credibility in these matters is everything, and nothing is easier to lose. One false alarm puts about 2.9 strikes against you for future discovery reports and the record is never expunged.

If your would-be discovery passes all these tests, then you're ready to make a report. Send an e-mail message to both marsden@cfa.harvard.edu and green@cfa.harvard.edu. E-mail? Yes. The telegram method Tom Bopp used to report his independent discovery of Comet Hale-Bopp no longer works — it was used too seldom to justify the expense of continuing it. The Central Bureau recommends reporting by e-mail, which has become the *de facto* postal system for astronomers all over the world.

Does hunting for comets sounds like a lot of drudgery for a pretty remote shot at immortality? Well, it's not for everyone but it does offer rewards. Even if you never find a comet of your own, you'll certainly become on first-name terms with a great many of the sky's wonders, and that's a hefty payback just in itself.

Good luck! And maybe the next great comet to be found will be yours.

Opposite. **If you're lucky and discover a comet, it will carry your name forever — which confers a kind of immortality. But many long hours of fruitless searching typically precede a discovery, and only the most dedicated hunters succeed. Even if you never find a comet, however, the search will give you a deeply rewarding familiarity with the starry sky. (Comet Hyakutake in a 4-minute exposure on Fuji SuperG 800 Plus print film using a Ceravolo 190mm f/2.3 Maksutov-Newtonian astrograph.)**

Orbital Elements for Comet Hale-Bopp

The following are the orbital elements for Comet Hale-Bopp, as determined by comet scientist Donald Yeomans of the Jet Propulsion Laboratory. He used more than 1250 observations to compile the elements. Many "sky viewing" programs that run on home computers let you add objects to their data base; just input these orbital elements in the format that your software requires.

Orbital Elements for Comet Hale-Bopp (1995 O1)

Corrected Elements (J2000): Solution 45
Epoch 2450520.50000 = 1997 Mar 13.00000

		Post-Fit Std.Dev.		
e	0.995095916	.000001646		
q	0.914101515	.000002361		
Tp	2450539.6346497	.0004084	1997 Apr	1.13465
Node	282.4706903	.0000059		
w	130.5909135	.0001366		
i	89.4294098	.0000462		

The following (J2000) osculating orbital elements can be used to generate ephemeris data using two body programs. However, care must be taken to select an orbital element set with an epoch close to the desired ephemeris output times.

Epoch (TDB)	e	q	Node	w	i	Tp	
1996 Oct 1.0	.9952307	.9143046	282.472419	130.575162	89.432510	1997 Apr	1.124093
1997 Jan 12.0	.9950918	.9141138	282.470878	130.589797	89.429555	1997 Apr	1.134271
1997 Mar 13.0	.9950959	.9141015	282.470690	130.590914	89.429410	1997 Apr	1.134650
1997 May 2.0	.9951088	.9141013	282.470806	130.591009	89.429283	1997 Apr	1.134718
1997 Sep 30.0	.9950814	.9140210	282.469240	130.585871	89.427986	1997 Apr	1.133652

e:	Eccentricity
q:	Perihelion passage distance (AU)
Node:	Longitude of the ascending node (deg.)
w:	Argument of perihelion (deg.)
i:	Inclination (deg.)
Tp:	Perihelion passage time (TDB)

Readings & Resources

To help you learn more about comets and comet-watching, it's important to get to know the resouces of your local public library and university library. (The latter will usually let you use the collection, even if you can't borrow anything.) Be sure also to check out the used book stores in your locality.

Just as important is getting on-line and exploring the Internet. This is especially vital for comets, because new discoveries can occur faster than the monthly magazines report on them. (Comet Hyakutake in March and April of 1996 was a perfect example.) If you have access through your home computer — great. But if you don't, all is not lost. More and more public and university libraries are providing access to on-line information. Terminals are often in the library's reference department; if you don't see them, ask.

Books on comets

Rendezvous in Space, by John C. Brandt and Robert D. Chapman, (W.H. Freeman, 1992). Subtitled "the science of comets," this book tells you what astronomers have learned about these objects, based on research including the recent return of Halley's Comet. Written for the interested lay reader.

The Mystery of Comets, by Fred L. Whipple, (Smithsonian Institution Press, 1985). Whipple authored the accurate "dirty snowball" model for comets almost 50 years ago. This personal account was written on the eve of the Halley flybys and is aimed at a more popular audience than the above title.

Observing Comets, Asteroids, Meteors, and the Zodiacal Light, by Stephen J. Edberg and David H. Levy, (Cambridge University Press, 1994). The section on observing comets is practical and thorough, and the rest of the book provides the same quality information. Aimed at amateur astronomers and observers.

The Quest for Comets, by David H. Levy, (Plenum, 1994). A detailed and personal look at the process of hunting for comets by an amateur/professional who has found many comets on his own and in collaboration with others.

Rain of Iron and Ice, by John S. Lewis, (Addison-Wesley, 1996). An excellent popular account of the dangers posed by asteroid and comet impacts. It covers the same topic as the next item below, but at a more general level.

Rogue Asteroids and Doomsday Comets, by Duncan Steel, (John Wiley & Sons, 1995). Ignore the tabloid-style title, this is a serious and absorbing book by a professional astronomer who is one of the few engaged in the detection of comets or asteroids that could strike Earth. A sober (and sobering) look by an insider at a real problem that's too often dismissed.

Comets, by Donald K. Yeomans, (John Wiley & Sons, 1991). This bears the subtitle "a chronological history of observation, science, myth, and folklore," which is certainly accurate. Nearly 500 pages long, it's an ideal source for looking up data on comets, or for simply reading about them.

Comet, by Carl Sagan and Ann Druyan, (Random House, 1985). This was published to catch the Comet Halley wave, but for readers today it also has great value in bringing together loads of colorful comet images (and not just of Halley): photos, drawings, period engravings and etchings, paintings — you name it.

Comets, a Descriptive Catalog, by Gary W. Kronk, (Enslow Publishers, 1984). Covering all comets discovered between 371 BC and AD 1982, this is a massive compendium of data on how a given comet was discovered, in what constellation, and what the apparition was like. Not something you would sit down to read continuously, but rather dip into.

Internet web-sites on comets

Comet observations home page:
http://encke.jpl.nasa.gov/

The Jet Propulsion Laboratory's main comet home page:
http://newproducts.jpl.nasa.gov/comet/

Sky & Telescope magazine's comet home page:
http://www.skypub.com/comets/comets.html

ASTRONOMY magazine's home page:
http://www.kalmbach.com/astro/astronomy.html

Comet and asteroid impact hazards
http://ccf.arc.nasa.gov/sst/main.html

Latest list of transneptunian objects
http://cfa-www.harvard.edu/cfa/ps/lists/TNOs.html

New comet discoveries are reported to the world through the quaintly named Central Bureau for Astronomical Telegrams, which is headquartered at the Smithsonian Astrophysical Observatory (see chapter 5). Despite its name, the "telegrams" arrive by e-mail. For information of the kinds of subscriptions offered, go to http://cfa-www.harvard.edu/cfa/ps/cbat.html or send a message to iausubs@cfa.harvard.edu asking for information.

CD-ROM

Comet Explorer, (Cyanogen Productions, 1996). A beautiful look at the apparition of Comet Hyakutake. It has stunning video sequences of the comet, both in wide-angle view and close-ups on the nucleus.

Books & periodicals about astronomy and the solar system

ASTRONOMY is a colorful and graphic monthly magazine devoted to presenting to the general reader the science of astronomy and backyard observing.

Sky & Telescope is also a monthly covering the same fields. It aims its stories towards the more experienced amateur astronomer and to the professional.

The Penguin Dictionary of Astronomy, by Jacqueline Mitton, (Penguin, 1993). How to tell an anti-tail from a nucleus, and other perplexing subjects.

General observing guides

Skywatching, by David H. Levy, (The Nature Company and Time-Life Books, 1995). An outstandingly informative and downright beautiful beginner's guide to all aspects of astronomy.

The Backyard Astronomer's Guide, by Terence Dickinson and Alan Dyer, (Camden House, 1991). Thorough and authoritative. Covers the entire field of amateur astronomy with solid, reliable information on what to do and provides clear-headed advice on equipment.

Star atlases

Sky Atlas 2000.0, by Wil Tirion, (Cambridge University Press, 1981). The best all-around star atlas and one that covers the whole sky. Useful for all levels of expertise.

Uranometria 2000.0, by Wil Tirion, Barry Rappaport, and George Lovi, (Willmann-Bell, 1987 & 1988). Covers both celestial hemispheres in two volumes, it's the atlas that every advanced amateur astronomer should own. Essential to help prevent false alarms in comet hunting.

Astrophotography

Astrophotography For the Amateur, by Michael A. Covington, (Cambridge University Press, 1991). An excellent first book to get if you want to try celestial picture-taking.

About the Solar System

Stardust to Planets, by Harry Y. McSween, Jr., (St. Martin's Press, 1993). A good survey of the solar system for armchair astronomers; engagingly written.

The Grand Tour, by Ron Miller and William K. Hartmann, (Workman Publishing, 1993). Color paintings and simple descriptions take you on a tour of the solar system.

The New Solar System, edited by J. Kelly Beatty and Andrew Chaikin, (Cambridge University Press, 1990). Now in its third edition, this contains *Scientific-American*-level chapters about every important solar system topic.

Moons and Planets, by William K. Hartmann, (Wadsworth, 1992). If you want to know how the solar system works, this is the best introductory textbook on the subject. Quite readable.

Introduction to Observing and Photographing the Solar System, by Thomas A. Dobbins, Donald C. Parker, and Charles F. Capen, (Willmann-Bell, 1988). A useful guidebook to exploring the Sun's family with an amateur telescope.

Software

Expert Astronomer (IBM & Macintosh) is an inexpensive and basic "sky viewing" program. It shows you the sky (using 9,100 stars) from any given location at a given date and time.

Dance of the Planets (IBM) is an impressive package that has excellent graphics and the ability to help you visualize the solar system from anywhere within it, on any date.

Starry Night (Macintosh) has the easiest-to-use set of controls of any sky view program. I used it in the preparation of this book; it's outstanding.

Voyager II (Macintosh) is a slick and powerful sky viewing program with over 50,000 stars and 4,200 deep-sky objects.

This is just the tip of an iceberg, and by no means mentions all the good programs. To stay current on what's available, check the issues of ASTRONOMY and *Sky & Telescope* (and their Internet web pages) for updates and reports on new software.

Organizations

Association of Lunar and Planetary Observers (ALPO), P.O. Box 143, Heber Springs, AR 72543, USA. An amateur group that carries out planetary patrol observations.

Astronomical Society of the Pacific (ASP), 390 Ashton Ave., San Francisco, CA 94112, USA. A national organization of amateur and professional astronomers.

International Dark-Sky Association, 3545 N. Stewart St., Tucson, AZ 85716, USA. Dedicated to helping astronomers of all kinds preserve dark skies for observing.

The Planetary Society, 65 N. Catalina Ave., Pasadena, CA 91106, USA. A space-advocacy group promoting solar system exploration by both unmanned spacecraft and human expeditions.

Royal Astronomical Society of Canada (RASC), 136 Dupont St., Toronto, Ontario, Canada M5R 1V2. The national organization for professional and amateur astronomers; has local "centres" (clubs) in major cities across Canada.

British Astronomical Association (BAA), Burlington House, Piccadilly, London W1V 9AG, UK. The national society for amateur astronomers.

To find out if there's a planetarium or astronomy club in your area, ask at the public library or a local college or university, or better still surf the world wide web!

Glossary

albedo. The fraction of light falling on a body that is reflected. The nucleus of Comet Halley has an albedo of only 0.03 — 3% — making it one of the darkest natural objects known.

anti-tail. An optical illusion produced when a comet's dust tail, lagging behind the comet in its orbital path, appears to point toward the Sun.

aphelion. The point in a comet's or a planet's orbit where it lies farthest from the Sun.

apparent magnitude. The brightness of a celestial object as seen from Earth. For comets (which are extended objects) it represents the magnitude the comet would have if all of it were concentrated into a starlike point.

apparition. The period lasting days, weeks, or months when a comet or other celestial object is best visible.

asteroid (minor planet). A rocky object orbiting the Sun and less than a thousand kilometers in diameter; some may be the nucleus of burned-out comets.

astronomical unit (AU). The average distance between the Earth and Sun, roughly 150 million kilometers or 93 million miles. (This is the handiest unit for measuring distances within the solar system.)

circumpolar objects. Celestial objects that never set as seen from a given latitude.

coma. The cloud of dusty gas surrounding the nucleus of a comet. Active when a comet is closer to the Sun than about 4 astronomical units, it effectively shrouds the nucleus from view and provides the source for the comet's gas and dust tails.

comet. A small body composed of ices and dust — a "dirty snowball" — which orbits the Sun on an elongated path. When near the Sun the nucleus heats up, shoots out jets of gas, develops a hazy gas coma, and streams off long tails of dust and gas.

dark adaptation. A process by which the human eye increases sensitivity under conditions of low illumination. Most of the change occurs within half an hour, but adaptation continues for several hours more.

degree. A unit of angular measure equal to one 360th of a full circle. Your forefinger held at arm's length is about 2° wide.

disconnection event. When a gust in the solar wind tears off part of the comet's gas tail.

dust tail. The streaming tail of a comet that is comprised of rocky dust shed by the comet's nucleus. Yellowish in color, it is driven away from the comet's coma by the pressure of sunlight. Often more prominent after a comet has passed perihelion.

ecliptic. The plane of Earth's orbit. A fundamental reference on the sky, it is the imaginary path followed by the Sun over the course of a year. Most planets' orbits lie close to the ecliptic.

gas tail. The plume of ionized gas that streams away from a comet's coma. Bluish in color (from ionized carbon monoxide), it is shaped by the charged particles in the solar wind and always points directly away from the Sun, even when the comet is returning to deep space.

Great Comet. A loosely defined term applied to any comet that is bright enough to become visible in daytime.

ion. Any atom or molecule that has added or lost an electron. The atom or molecule then becomes electrically charged. A comet's gas tail contains ionized gases.

Kuiper Belt. A region of comets extending from 35 astronomical units to about 1,000 AU. Believed to be a relatively flat disk in the plane of the solar system, it is also the probable source for short-period comets. Its namesake was Gerard Kuiper (1905-1973), a Dutch-American astronomer who first proposed its existence.

light-year. The distance that light travels in one year: 9.5 trillion kilometers.

long-period comet. By arbitrary convention, any comet with an orbital period greater than 200 years.

magnitude. A logarithmically based unit used to measure the optical brightness of celestial objects. For historical reasons, numerically lower magnitudes (which can run into negative numbers) are brighter than numerically larger magnitudes. A 5-magnitude difference represents a 100-fold change in brightness.

magnitude, absolute. For comets, the brightness it would have at 1 astronomical unit from both the Sun and Earth. (In the case of a star, it's how bright it would be if placed at a distance of 10 parsecs, or 32.6 light-years.)

meteor (shooting star). The bright transient streak of light produced by a bit of space debris burning up as it enters the atmosphere at high speed. Most are tiny pieces of rocky dust from comets, about as friable as cigar ash.

meteor shower. A large number of meteors that appear to come from one small region of sky. It is produced when Earth encounters the debris thrown off by a comet. Meteor showers are named for the constellation from which they appear to come.

nongravitational effects. Changes in a comet's orbit brought about by the jets of the comet's nucleus acting as weak rocket thrusters.

nucleus. The solid head of a comet, composed of ices and dust, and believed to be a few miles or kilometers in size. Whipple coined the term "dirty snowball" as a description and it has stuck.

Oort Cloud. A cloud of comets surrounding the solar system named for Jan Oort (1900-1992), the Dutch astronomer who first proposed it. The main Oort Cloud extends from about 20,000 astronomical units out to some 100,000 AU. Its shape is generally spherical and the outer edge is only loosely defined. There is also an inner Oort Cloud extending inward from about 20,000 AU. Its inner edge becomes disk-like and lies in the plane of the solar system, merging into the Kuiper Belt at around 1,000 AU.

orbit. The path followed by any celestial object moving under the control of another's gravity.

perihelion. The point in the orbit of a planet or a comet where it lies closest to the Sun.

perturbation. A change in the orbit of one body caused by the gravitational pull of another.

planetesimal. A smaller body that may collide with others to form a planet, or a large fragment from a planet-shattering collision.

radiation pressure. The outward force exerted by sunlight on dust particles in a cometary coma and tail.

seeing. A measure of the steadiness of the atmosphere; good seeing is essential to using high magnifications.

short-period comets. By convention, comets with orbital periods less than 200 years.

solar nebula. A disk of dust and gas surrounding the newborn Sun; out of it were born the planets, asteroids, and comets.

solar system. Everything that orbits the Sun: nine planets (plus their satellites), thousands of asteroids, and countless comets, meteors, and other debris.

solar wind. A high-speed stream of particles blowing from the Sun in all directions. It carries a magnetic field and interacts will all electrically-charged matter (such as the ionized gases in a comet's head).

tail. The part of a comet that streams away from the nucleus, made of dust and gas. See *dust tail* and *gas tail*.

transneptunian objects. Objects that are relatively large (100 miles' diameter), dark, and reddish in color. They orbit from about 30 astronomical units out into the Kuiper Belt and may be the source for new short-period comets.

Universal Time. The time reckoning used by astronomers — essentially the same as Greenwich Mean Time. It is exactly 5 hours later than Eastern Standard Time.